Note: The following activities can be done with any of the units or with units you develop on your ow

General Ideas and Activities

Field Trip: Take class trips to local industries. These trips do not have to be to huge, major companies. A trip to a small business that uses a simplified assembly line of people and equipment can be easier for children to understand.

1. Pre-planning — Decide where you are going and make arrangements well ahead of the actual trip.

   Prepare children by discussing what they need to watch for during their visit. Work with the students to decide on questions they wish to have answered on the trip. Make different children (or groups) responsible for finding out the answers to specific questions.

   Be sure you have film in your camera and that the batteries work. A picture record is especially helpful when visiting a factory.

2. Trip — Be sure you have plenty of adult supervision! Review guidelines on how to act responsibly on a field trip. Remind your students and adult helpers of the importance of safety.

3. Post-trip — Provide ample time for discussing what was seen. Let each group answer the question they were responsible for (with input from the rest of the class).

   Share the pictures from the trip on a bulletin board or in a photo album. You may want to include the questions and answers from the trip.

   Write thank-you letters to the appropriate people.

   Prepare drawn and/or written records of the trip. A large mural is a good way to incorporate all the steps in a manufacturing process.

Speakers : Invite people from industries in your own community to talk to your class about their business, how it functions, and what their specific job is. Don't overlook the parents of your students. They can frequently be a valuable source of information.

Welcome to Our Class

Assembly Line Skit: Use the skit on pages 3 and 4 to demonstrate how an assembly line works. There are parts for five children at a time. Two variations are also given. The format of the skit is simple enough that older children should be able to create other skits of their own. Maybe even one to go with sneakers, ice cream, or pencils.

Other Sources of Information:

Check your audio-visual catalog for movies or videos.

Write for information. Many companies have packets of information suitable for school children. Your students can practice their letter-writing skills as they collect valuable information. The research librarian at the public library will be able to help you find addresses of companies to write.

Assembly Line — Greeting Cards

by Leslie Tryon

In this skit the children act out the steps of an assembly line. They will be creating a greeting card. You may wish to use a seasonal design for the rubber stamp portion. The children will be able to use the finished cards. Use a colorful sheet of paper or tag 5½" X 8½" (14 X 21.6 cm).

There are five children involved in this skit. The steps in the assembly line are:

Child 1 — Fold the paper in half and pass it on.

Child 2 — Round the corners on the unfolded edge and pass it on.

Child 3 — Place the card with the fold to the left. Stamp with a rubber stamp in the center of the folded piece of paper. Pass it on.

Child 4 — Using a crayon or a marker, do a design around the edge of the card. This could be a solid line, dots, or dashes. Pass it on.

Child 5 — This child is quality control. He/She should look for a nice fold, the correct corners rounded, the rubber-stamped design right in the center, and a consistent border design.

(Child 6) You may wish to add these two additional parts: a child to hand the paper to the first child and a child to box the finished cards. (Child 7)

Character	Dialogue
Child 1	On this assembly line I fold the card. My job is fun and it isn't very hard.
Child 2	When he/she hands it to me, that's my cue, To round the corners, so that's what I do.
Child 3	I put the fold on the left, and then I stamp. Try not to touch it, it could still be damp. (Child recites the second line while passing the card to the next person.)
Child 4	I'll use a bright color and do a fancy border. Maybe some dots or some dashes are in order.
Child 5	I check each card. I do quality control. You might say I'm the detail patrol.

Variations for the assembly line skit:

Assembly Line — Wrap an Apple

Character	Dialogue
Child 1	I cut the tissue into a nice big square, Just the right size for an apple or a pear.
Child 2	I place the apple in the center just like that. I wrap it up and pinch it at the top like a hat.
Child 3	I'll tie it with a ribbon. I think I'll use blue. Now it's ready for a tag, so I'll hand it to you.
Child 4	I punch a hole in the tag, then I'm ready to go. I pull the ribbon through the hole, then I tie a bow.
Child 5	I check each apple. I'm the quality control. You might say that I'm the detail patrol.

Assembly Line — Fill a Jar

Character	Dialogue
Child 1	I make sure the jar is clean. I inspect it carefully. I want it to be clean. It's very important to me.
Child 2	I pour the liquid into the jar. I'm careful not to fill it up too far. (You may substitute the name of an actual liquid. For example: orange juice, tea, lemonade.)
Child 3	I screw the lid on nice and tight. I make sure that I do it just right.
Child 4	I put the label on the jar with lots of glue. And carefully pass the jar to you.
Child 5	I check each jar. I do quality control. You might say that I'm the detail patrol.

Sneakers

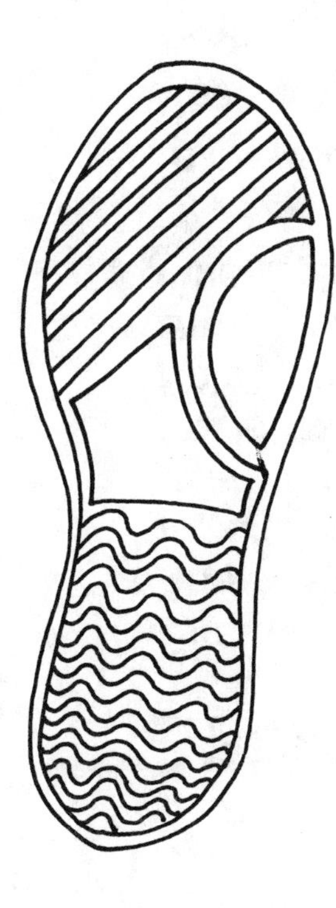

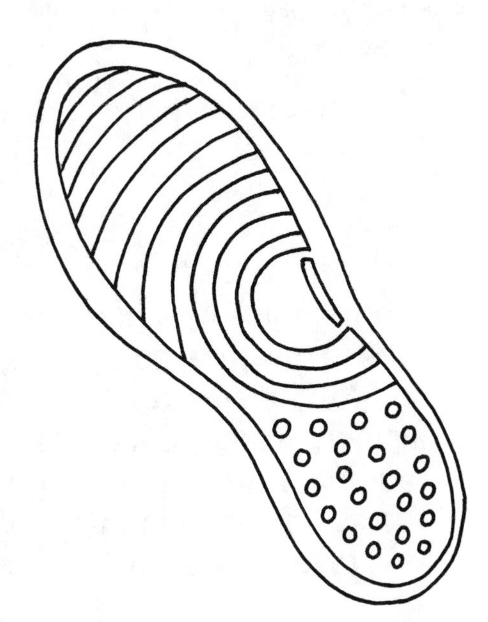

Rubber soles? I understand,
They're there so I won't slip.
Lots of lace? I understand,
They're there so I won't trip.

Nice high tops? I understand,
They're there so I won't wobble.
Arch supports? I understand,
They're there so I won't hobble.

Red or blue? I understand,
These match my running suit.
Stripes or dots? I understand,
They're fun, they're weird, they're cute.

But "sneakers"? I don't understand.
Were they really designed for sneaking?
Basketball sneaking? Or tennis sneaking?
It confuses me, in a manner of speaking.

by Leslie Tryon

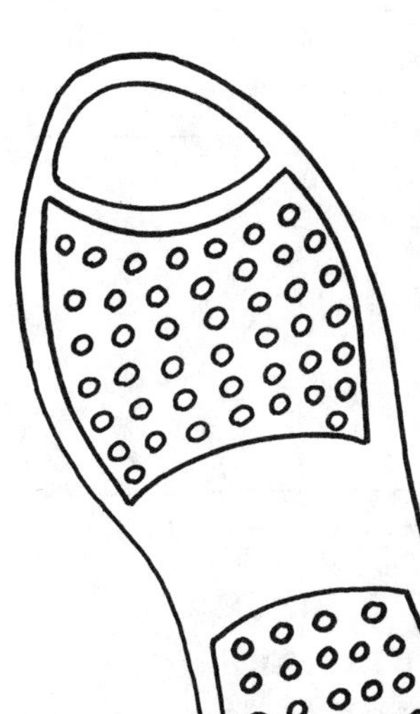

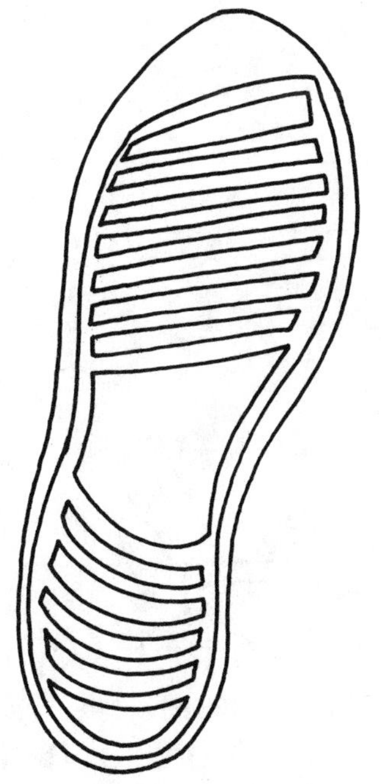

Books to Read about Sneakers (and other footwear):

Making Sneakers by Bruce McMillan; Houghton Mifflin Co., 1980

Shoes for Everyone, A Story about Jan Matzelinger by Barbara Mitchell; Carolrhoda Books, 1986

Sneakers by Samuel Americus Walker; Scholastic Book Services, 1978 (for teacher reference and older students)

Sneakers Meet Your Feet by Vicki Cobb; Little, Brown and Company, 1985

Note: Reproduce pages 6-8, fold in half and staple on the left to create a booklet to take home.

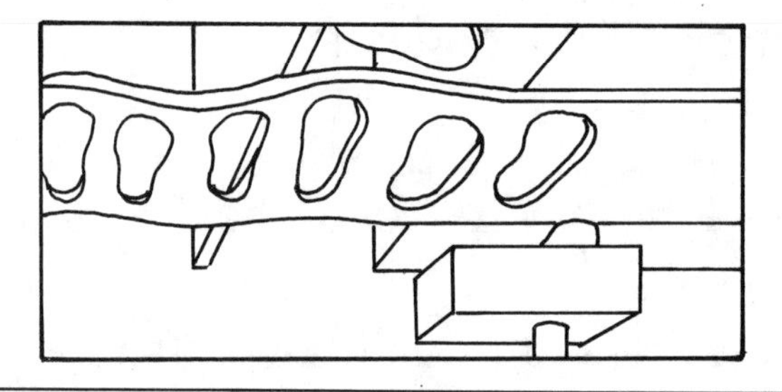

Let's see how a sneaker is made.

First the factory must get all the materials needed for the sneaker. The sneaker factory buys soles from a place that manufactures soles for shoes. The outside soles for sneakers are made of rubber. Rubber used to be made from the sap of a rubber tree. Now rubber is usually made from petroleum. The rubber is rolled out flat with special rollers. Soles are cut out of the rubber, put into molds and baked in ovens. Several different layers of material may be put together to make the inner sole of your sneakers.

Think About It:
What are you wearing that is made of rubber?
Make a list of other things that are made of rubber.

2

Fold

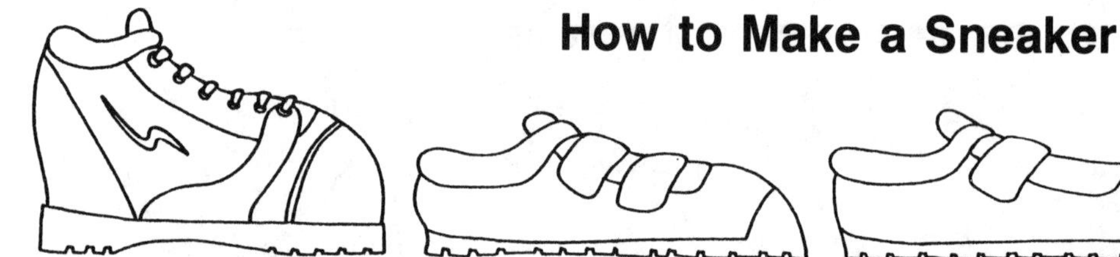
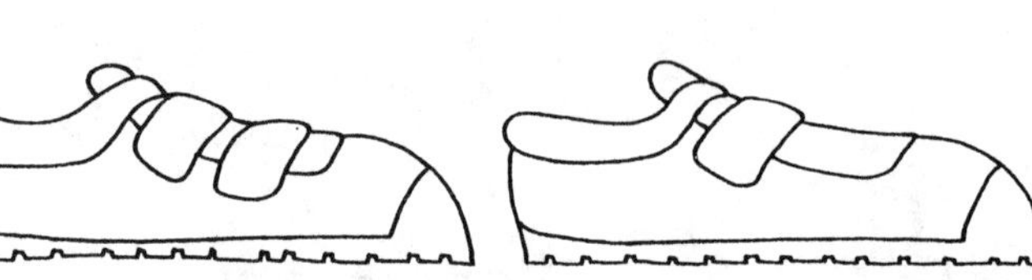

How to Make a Sneaker

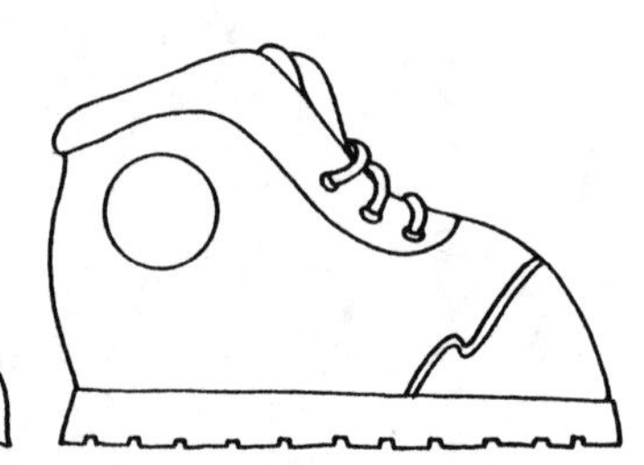

It isn't easy to make a pair of shoes. Long ago shoes were made by hand. Each pair could take the shoemaker several days to complete. A pair of hand-made shoes cost a lot of money, so they had to last a long time.

Now shoes are made in a factory. There are special machines to make the job faster. Many people work on one pair of shoes. Each person does one job. This is called an assembly line. This way several hundred pairs of shoes may be made in one day.

Think About It:
How do you think a shoe is made?
Make a list of other things that might be made on an assembly line.

1

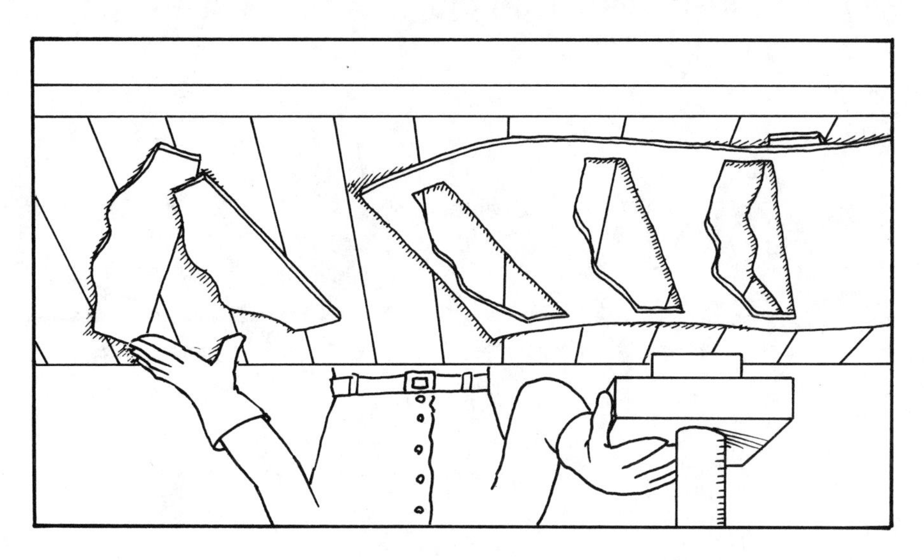

At the sneaker factory, cutters using special tools cut out the pieces for the upper part of the sneaker. The die looks like a cookie cutter. It is laid on the material. The press pushes the die through the material. The pieces are then sent on to the sewers.

Think About It:

How is making a pair of sneakers like baking cookies?

Fold

The material used for the tops of the sneakers may be canvas, leather, or nylon. Each of these comes from a different factory.

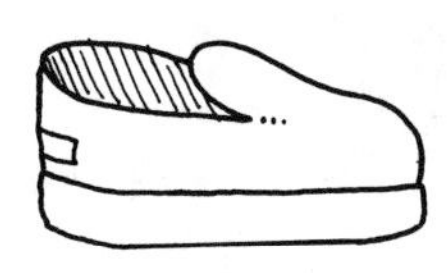

Canvas is made from heavy cotton cloth. The cotton cloth is woven from cotton thread. This thread is made from the fibers of the cotton plant.

Leather is made from animal hides. The animal skins must be specially treated before they can be used for leather shoes.

Nylon is made from petroleum. Chemicals in the petroleum are mixed together and squeezed through small holes to make threads. The threads are woven together to make nylon cloth.

Think About It:

Are you wearing anything made of cotton?
Are you wearing anything made of nylon?
What else can you name that is made from each of these materials?

The finished sneakers are put into boxes and sent to a warehouse. The shoes are kept in the warehouse until they are ordered by a store. Then they are put on a truck and delivered to the store where you can go to buy a brand new pair.

Think About It:

See if you can remember all of the steps it takes to make a sneaker. Why are you an important part of the sneaker industry?

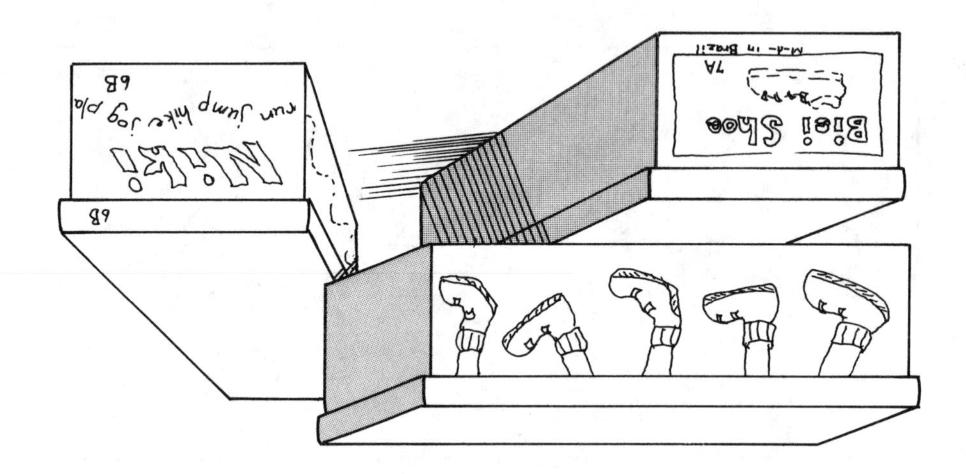

Fold

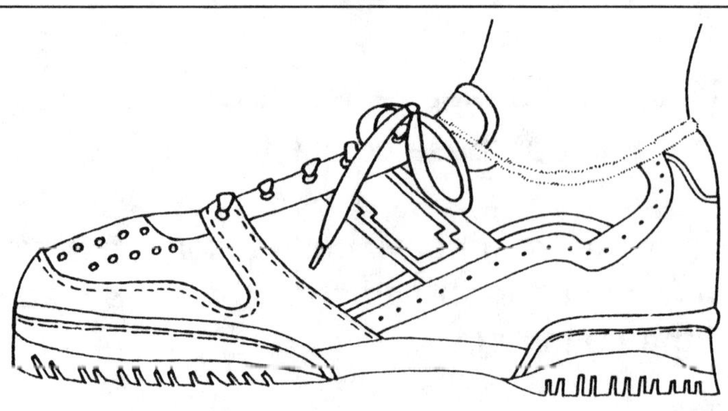

In the sewing room each worker does a different sewing job. The pieces move along the assembly line from worker to worker until all the pieces are sewn together.

The upper part of the sneaker is attached to the sole with a special glue. The finished sneaker is trimmed and cleaned and the shoelaces are put in. An inspector checks the finished shoes to be sure they have been made correctly.

Think About It:

What does the inspector need to check?

Note: These questions can be used in a whole class discussion or in cooperative learning groups.

Think About Sneakers

Why do people wear shoes at all?

Why are sneakers so popular?

How do you think sneakers got their name?

Make a list of other names sneakers can be called.

Sneakers come in pairs. What else do we buy in pairs?

Why do sneakers cost so much?

Can some sneakers be bad for your feet? Give reasons for your answer.

Which kind of sneakers are best—leather, canvas, or nylon?

Can you think of places where you should not wear sneakers?

Crazy Feet

Design sneakers for these folks:

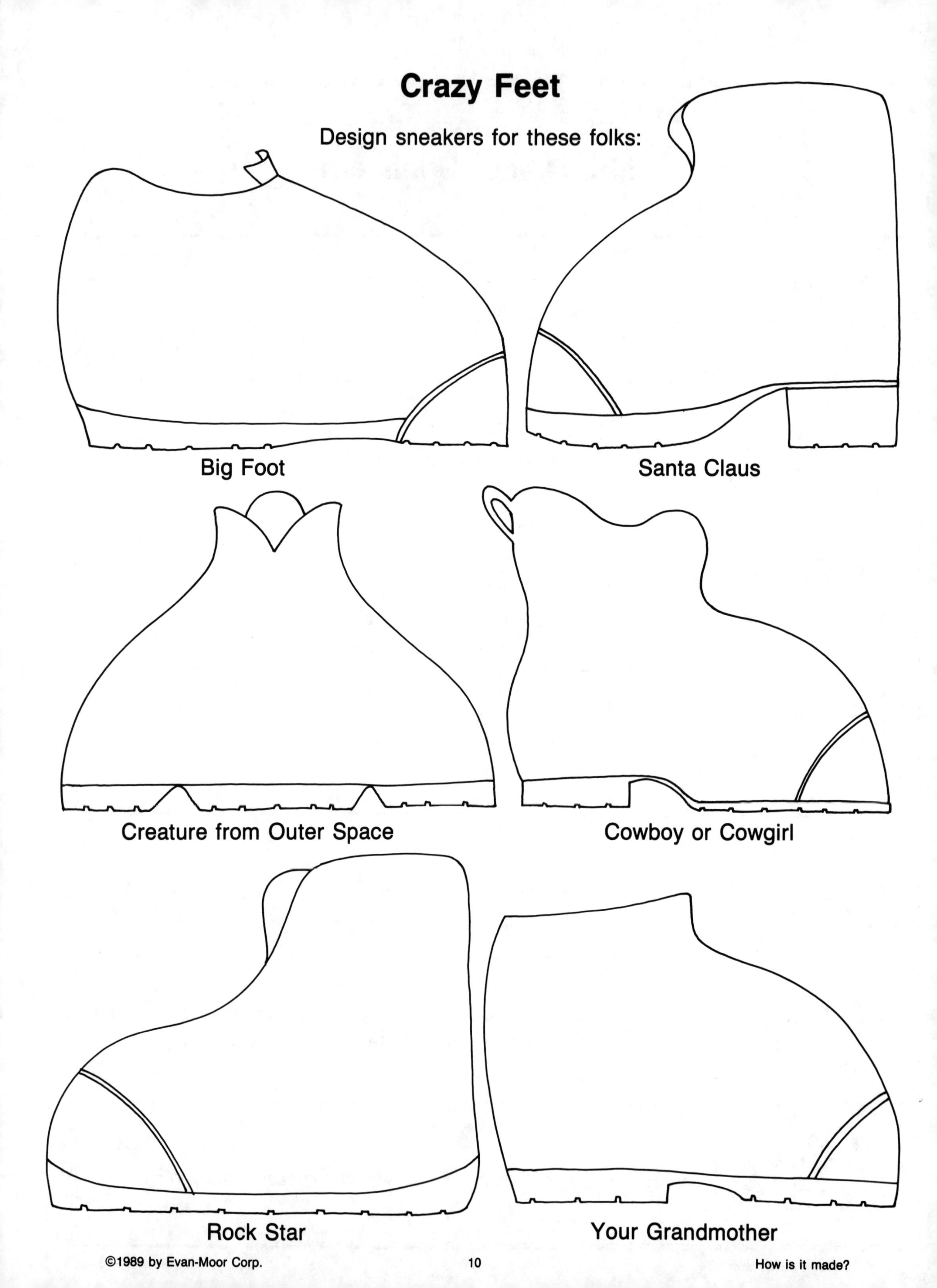

The Runner

Sometimes it pays to be quick on your feet.

Fill in the bubbles to explain WHO this runner is, WHERE the runner is going, and WHY it is important to get there quickly.

(You may draw different clothes to make the runner fit your story.)

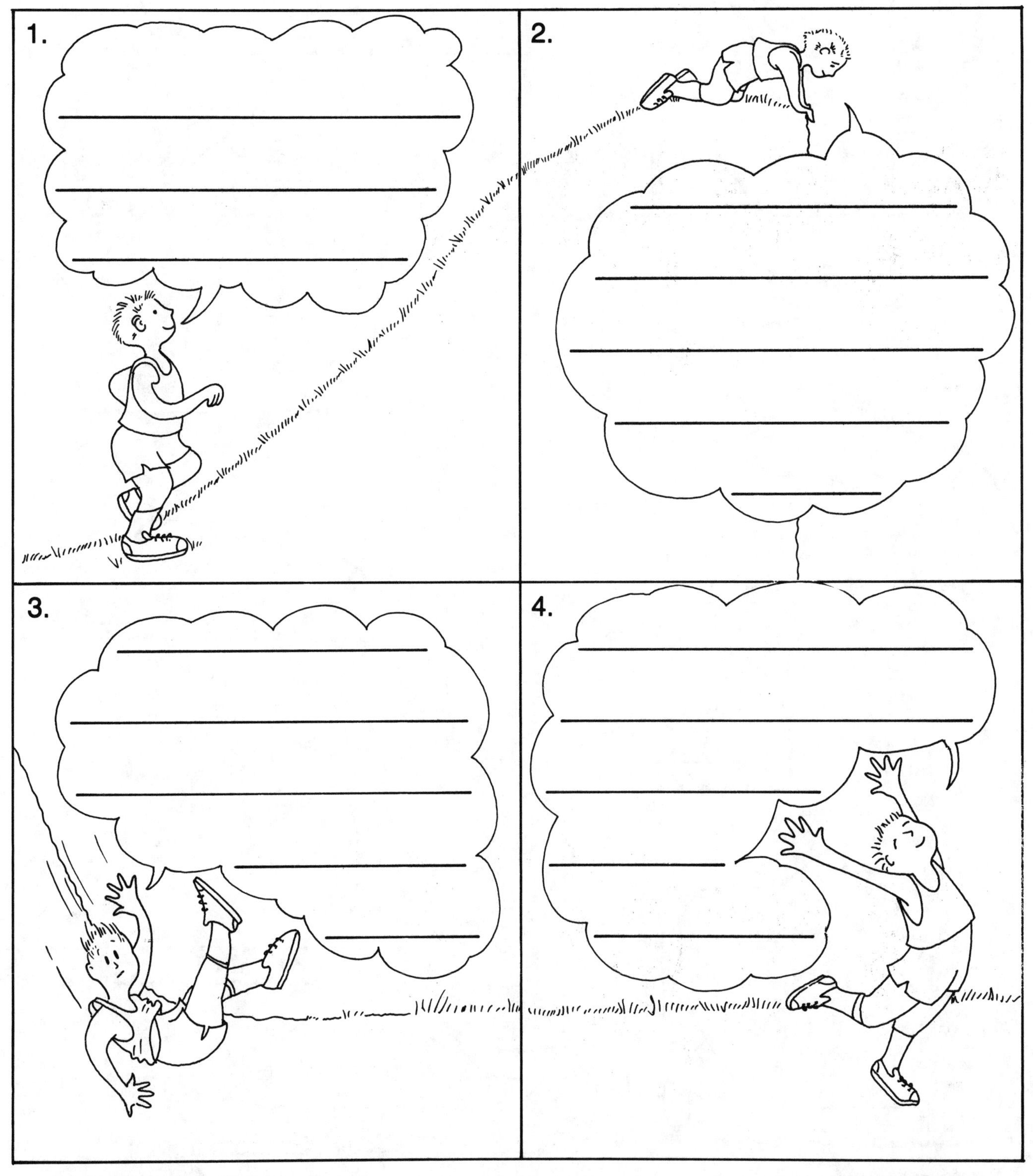

Match the Pairs of Sneakers

There was an accident in the sneaker shop.
All of the sneakers are mixed up.
Find each pair that match and make them the same color.

Note: Guide younger students through the steps of this activity. It can be done by individuals, in pairs, or in groups of four.

Designer Sneakers

There are dozens of types of sneakers for sale. Everyone has an idea of what the best sneaker should be like. Pretend you are a designer. Create a new sneaker for kids.

The name of my sneaker company is________________________.

My sneaker is make out of ___________________________.

This sneaker is great for ____________________________.

____

____

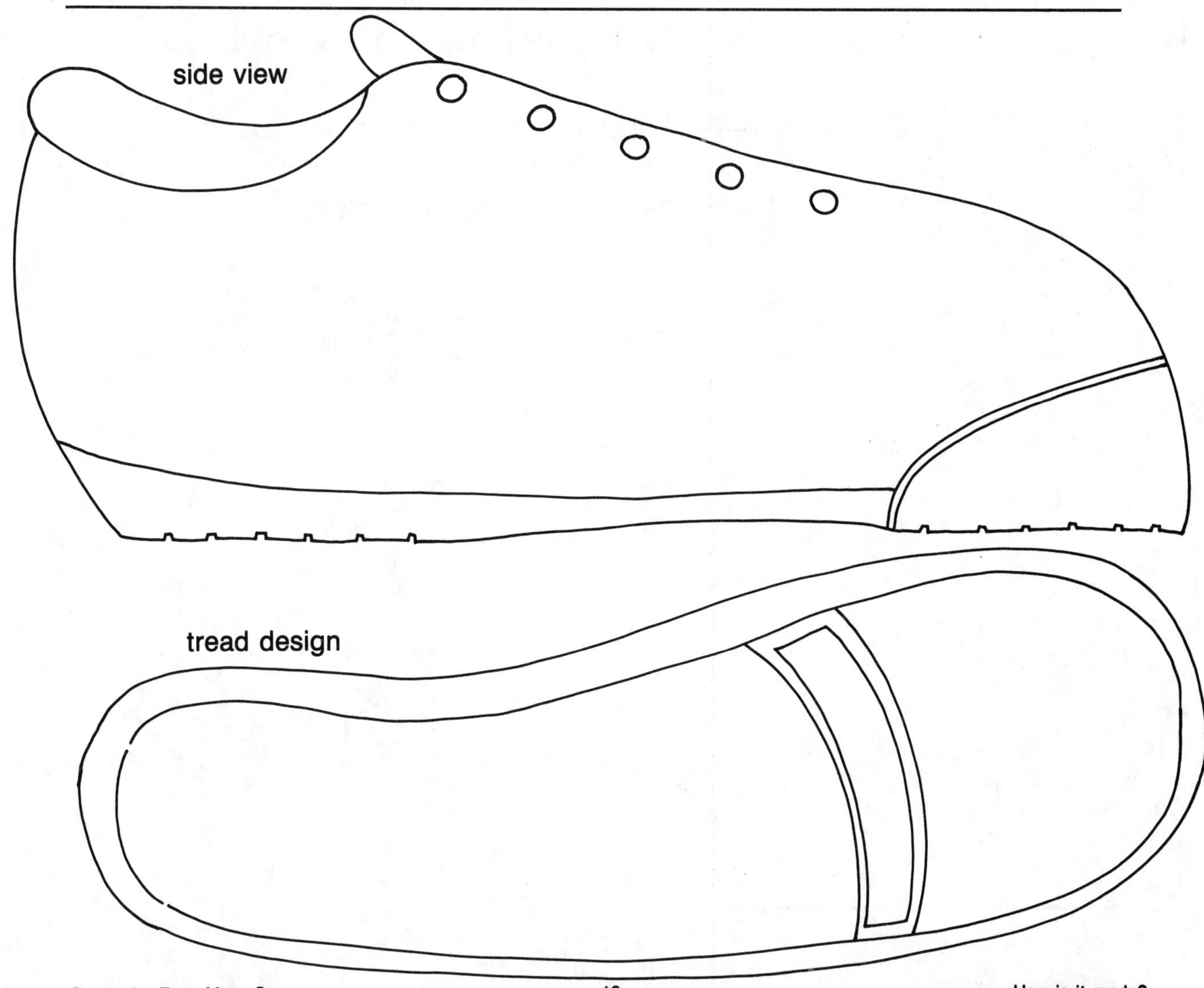

Note: Reproduce this pattern on construction paper. Each child will need scissors, crayons or marking pens, and yarn or string for a shoe lace. Have a hole-punch available also.

A Three-Dimensional Sneaker

Fold

1. Draw and color the design on your sneaker.
2. Cut it out and punch holes for the shoe lace.
3. Fold the sides up and lace your shoe.

Fold

Note: Have children use this pattern to cut out two construction paper sneakers as a cover and as many pieces of lined paper as they need for their stories.

A Sneaker Shape Book

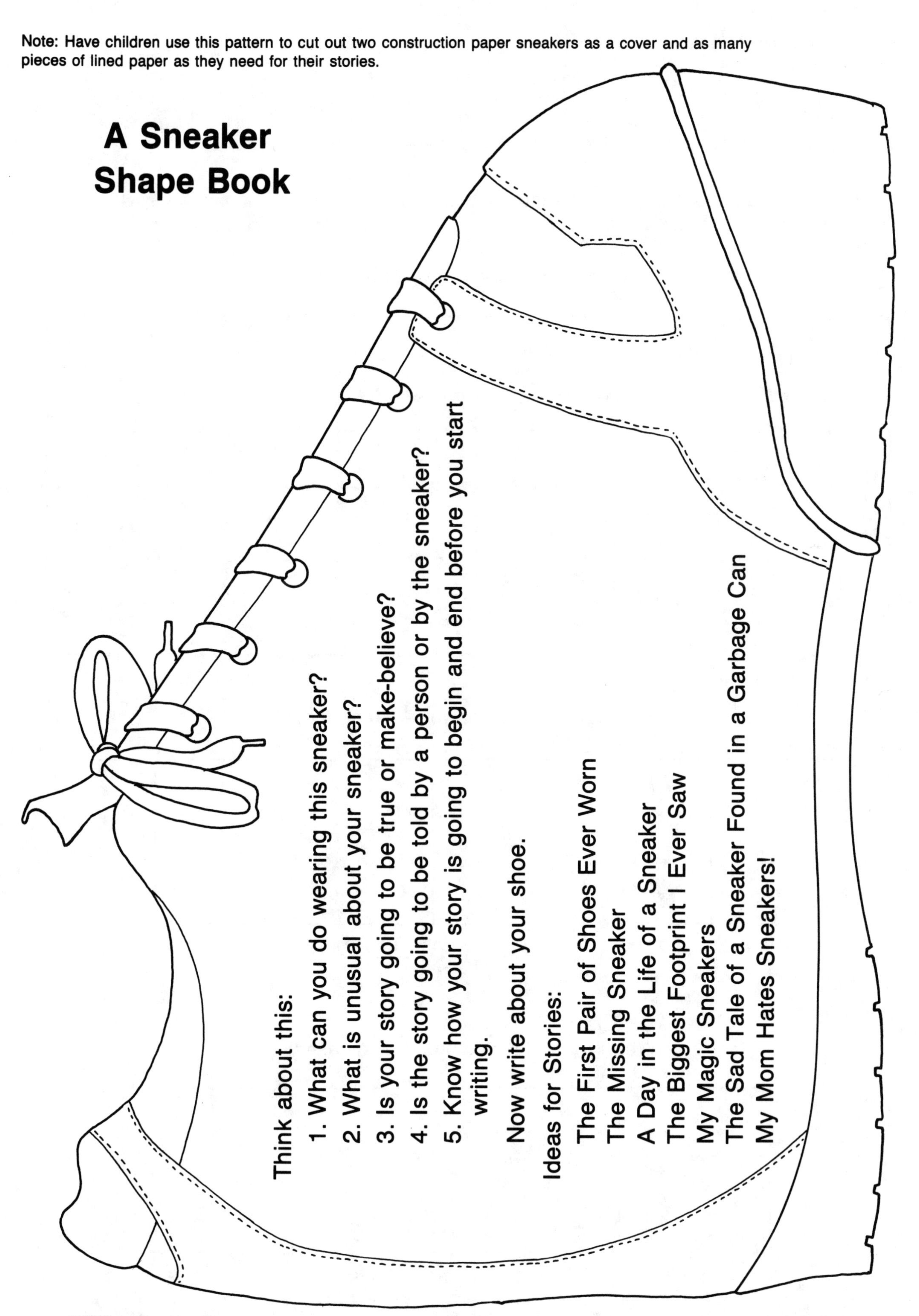

Note: Use sneakers to practice graphing in a variety of ways.

Sneaker Math

Ask a question such as those listed below and tally the answers.

Have your students record the information on a bar graph. This may be made on the chalkboard, on a sheet of butcher paper, or on an overhead projector.

Ask questions requiring children to read the graph to find the answers.

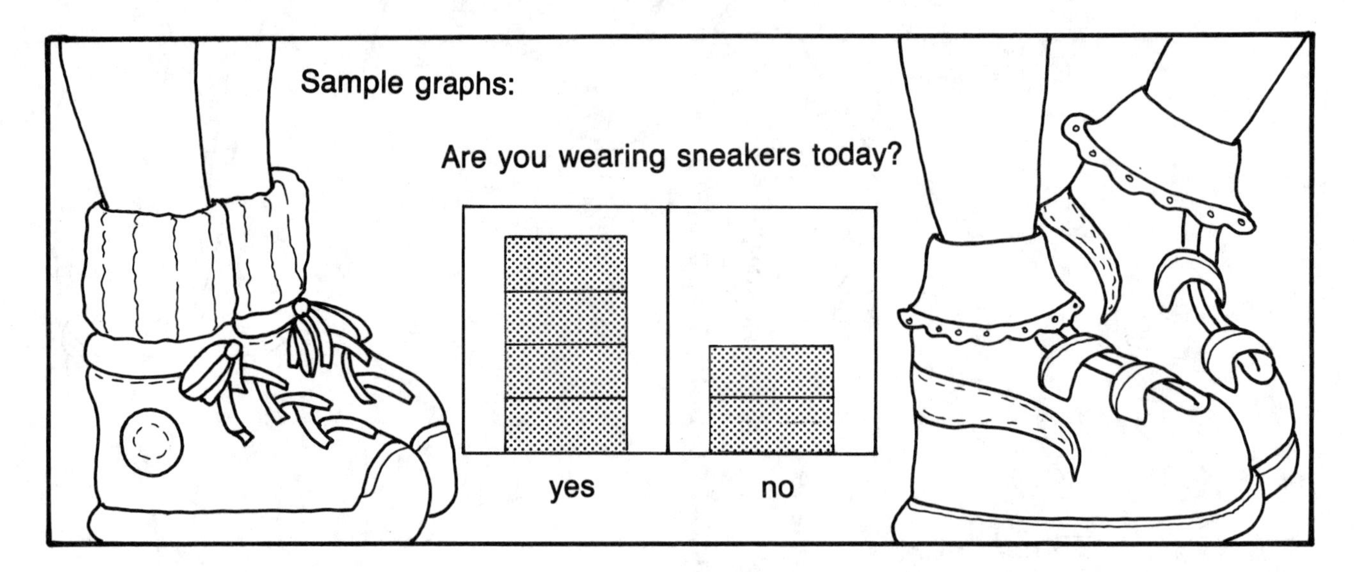

What type of shoe are you wearing today?

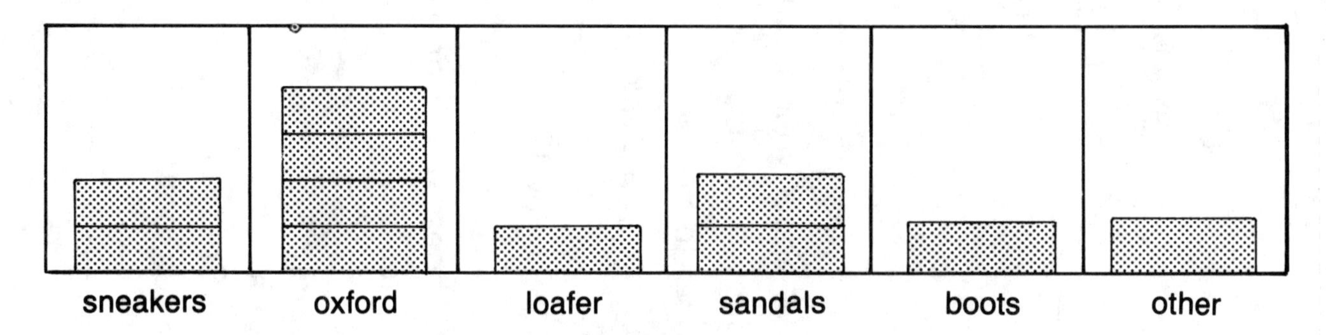

What is your favorite brand of sneaker?

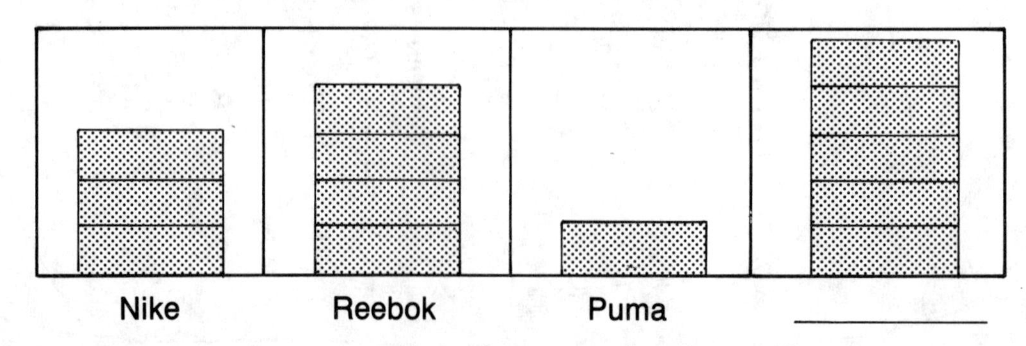

A Bulletin Board

Create this bulletin board to display your students' work.

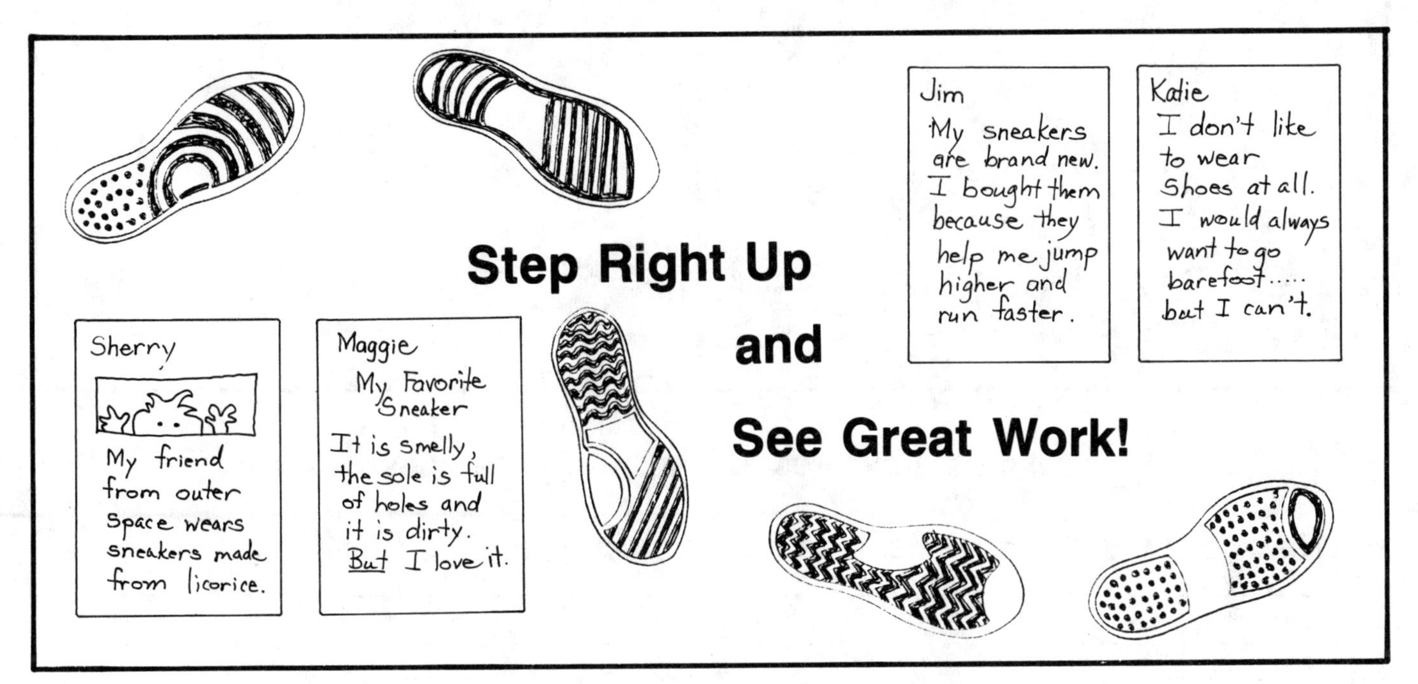

1. Back the bulletin board with dark blue butcher paper.
2. Add a sign made on a sheet of tag in a contrasting color.
3. Have children make "tread prints" using old sneakers and tempera paint directly on the blue butcher paper.

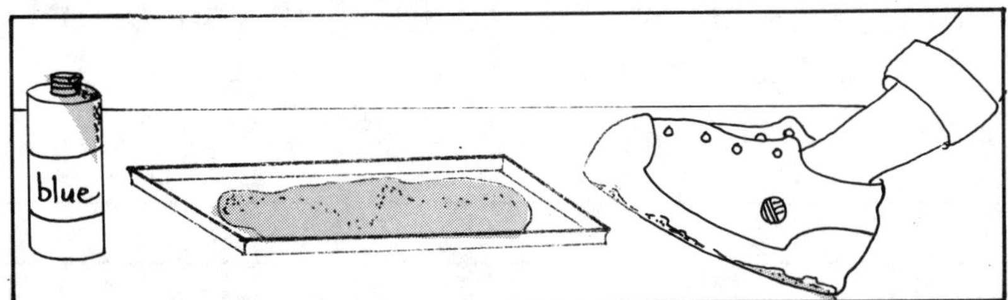

4. Pin up samples of your students' work.

Variations — Simply change the sign to fit a specific need. For example:

Step Right Up and Learn About Sneakers (Display facts about the history or construction of sneakers.)

Step Right Up and Read Our Stories (Display original stories about sneakers or other footwear.)

Step Right Up and Meet ______________________________
(Use the board to highlight a special student for the week. Have the other students place stories and pictures about the student on the bulletin board or allow the student being highlighted to use the board to display items about himself/herself — pictures, hobbies, talents, etc.)

Side view of a sneaker

Side view of the foot

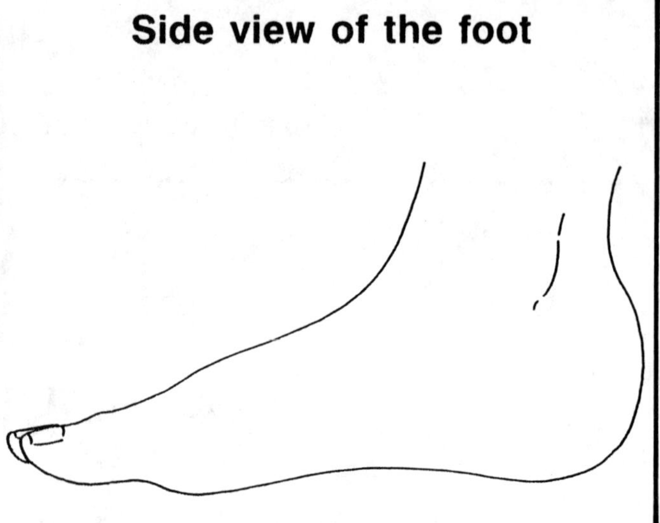

Look at the pictures and answer these questions:

1. How does the shape of the shoe help protect your feet?

2. Which parts of the shoe are shaped to match the shape of your foot?

3. What changes would you make in this sneaker to make it a better fit for the foot?

Other Sneaker Activities

Sneaker Day— Have everyone wear their sneakers (old, new, borrowed, or blue) to class for the day. Celebrate by making cookies shaped like the sole of a sneaker. Let everyone decorate the "tread" on their cookie. Have milk or juice on hand to wash the cookies down.

Mystery Sneakers— Go around your class (or school) and take pictures of feet in sneakers. Display the sneaker pictures and see if your students can identify who is wearing the sneakers. (Be sure to include yourself and other people on the staff.)

Draw Sneakers — Have your students take off one of their sneakers and draw it from the side, front, top, and/or bottom. Have an extra or two on hand for anyone not wearing sneakers that day.

Write Shape Poems— Guide your students through these steps:

1. Brainstorm to create a list of words and phrases that describe their favorite sneaker. List these on a chart or the chalkboard.

2. Have students select the words and phrases they like best and arrange them in a pleasing order. For example...

   torn and dirty
   laces gone
   sole torn loose
   my old sneakers are like a friend to me
   they are always there when I need them

3. Have your students draw a simple sneaker shape using a black crayon or marking pen. Clip a sheet of thin paper (typing paper works well) over the drawing. Write the poem following the shape of the sneaker. Remove the drawing and mount the shape poem on a sheet of colored paper.

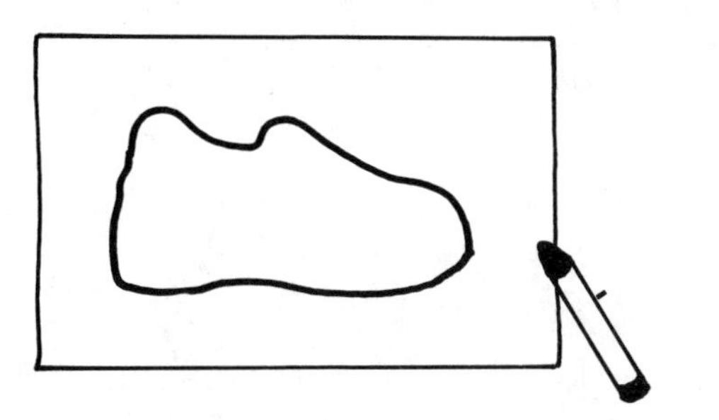

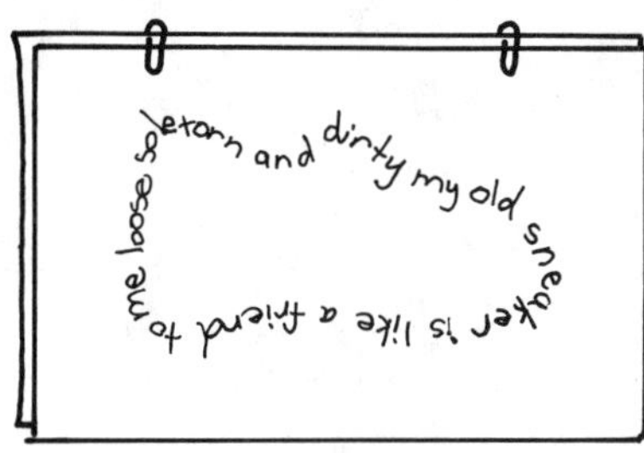

Rewrite — Older students can work in a group to update the story *The Shoemaker and the Elves* making the story take place in a small sneaker plant. Rewrite it as a modern day fairy tale or in play form to present to the class.

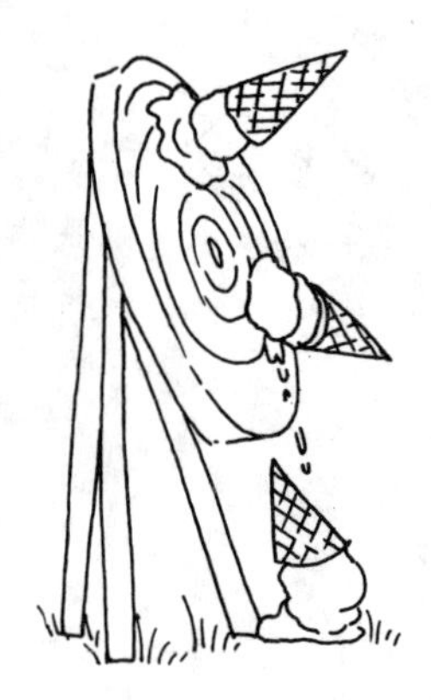

Ice Cream

Too many ice cream scoops — droop.
Ice cream cones that begin to drip — slip.
Ice cream cones, at the point or the peak — leak.
To eat ice cream on a very hot day you must lick — quick.
Ice cream in big glass dishes — delicious.
The one who stole my banana split — split.

by Leslie Tryon

Ice cream, ice cream
Cool and sweet
Ice cream, ice cream
My favorite treat.

by J.E. Moore

I scream, you scream
We all scream
For ice cream.

Folk Rhyme

Books to Read About Ice Cream:

From Milk to Ice Cream by Ali Mitgutsch; Carolrhoda Books, 1981
Ice Cream by William Jaspersohn; Macmillan, 1988
The Scoop on Ice Cream by Vicki Cobb; Little, Brown and Company, 1985
What Was it Before it Was Ice Cream? by Coleen Reece; Childs World, 1985

Note: Reproduce pages 21 to 24. Fold the pages in half. Staple the pages together on the left side to create a booklet children can take home.

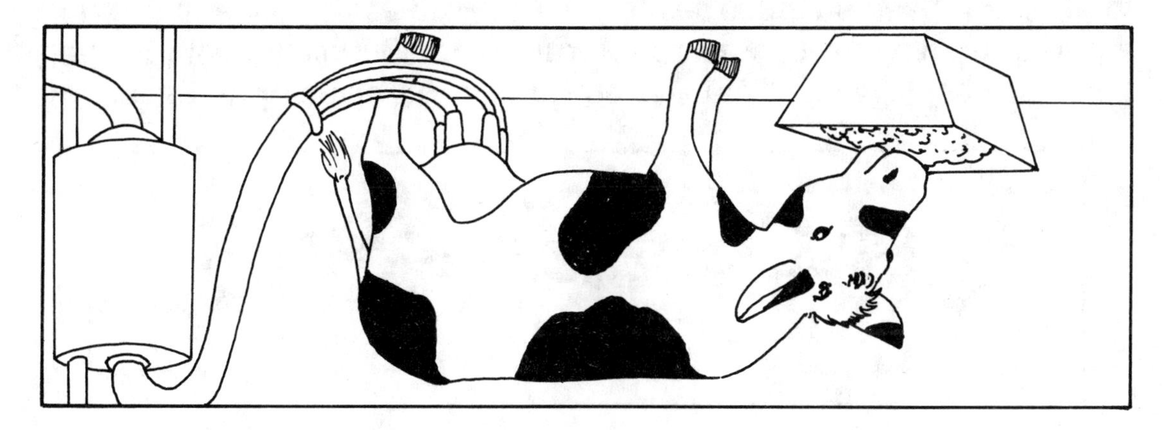

The milk or cream in ice cream comes from a dairy. Dairy farmers raise cows to get the milk. A cow can produce much more milk than she needs to feed her calf. The dairy farmer uses special machines to milk the cows. The milk is kept very cold so it will not spoil. The milk is carried to the ice cream factory in special tankers.

Think About It:

Can you name any other animal that produces milk used by people? Is the milk drunk or used in another way?

Fold

All About Ice Cream

Different kinds of ice cream contain different ingredients, but all ice cream contains milk in some form, sugar, and a flavoring. If ice cream is going to be stored, it must also contain stabilizers to keep it smooth.

Some of the ingredients in ice cream come from farms in the United States. Other ingredients come from faraway countries. Let's find out about these ingredients.

Think About It:

How many different flavors of ice cream can you name? How many of these flavors have you tasted?

Sugar comes from two special plants. One is a tall kind of grass called sugar cane that grows in hot parts of the world. (The other plant is the sugar beet.) The sugar cane is cut down and sent to a mill where the sweet juice is crushed out of the cane. The juice is boiled to remove the water. Sugar crystals covered in molasses are left. The sugar is sent to a refinery where the sugar is turned into a syrup and the molasses is removed. It then goes through filters to remove any impurities. When the water is taken out, the sugar looks like the sugar you use at home. Some of the sugar syrup goes into special tankers to be taken to the ice cream factory.

sugar beet

sugarcane

Think About It:

Take a magnifying glass and some sugar. Can you see why sugar is called a crystal? Can you name another crystal that people eat every day?

3

Fold

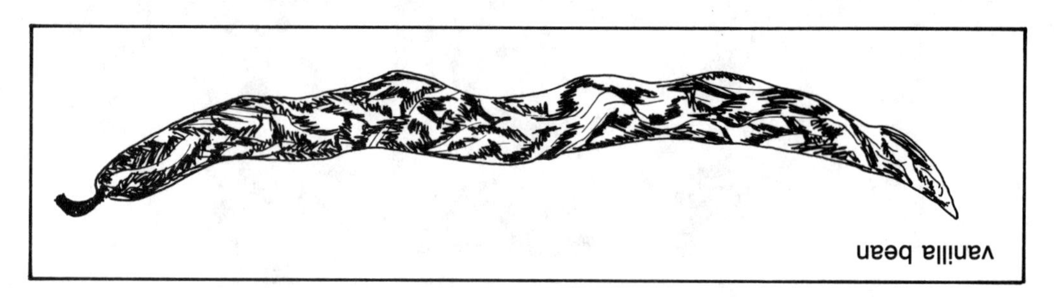

Chocolate and vanilla are two of the favorite flavors of ice cream. Both of these flavorings come from far away. In tropical parts of the world, a special flower grows. This flower contains seed pods full of tiny black seeds. These seed pods are used to make vanilla extract. The vanilla beans are dried in the sun. Then the beans are sent to a factory where the beans are chopped and put into large tanks. A mixture of alcohol and water is poured through the chopped beans. The vanilla dissolves in the liquid. It is stored in kegs for several months before it is delivered to the ice cream plant.

Think About It:

See how many flavorings you can name that are used at your house in cooking the food you eat.

4

At the ice cream factory, milk, liquid sugar, and stabilizers are mixed together. The mix master's job is to make sure that the finished ice cream is always the same by adjusting the amount of each ingredient being used. The mixture is heated then cooled quickly to kill any germs that might be in it. The ice cream must be frozen quickly and must be beaten. This keeps the ice crystals small and whips air into the mixture. This makes the ice cream smooth and creamy. Now the mixture looks like a thick paste.

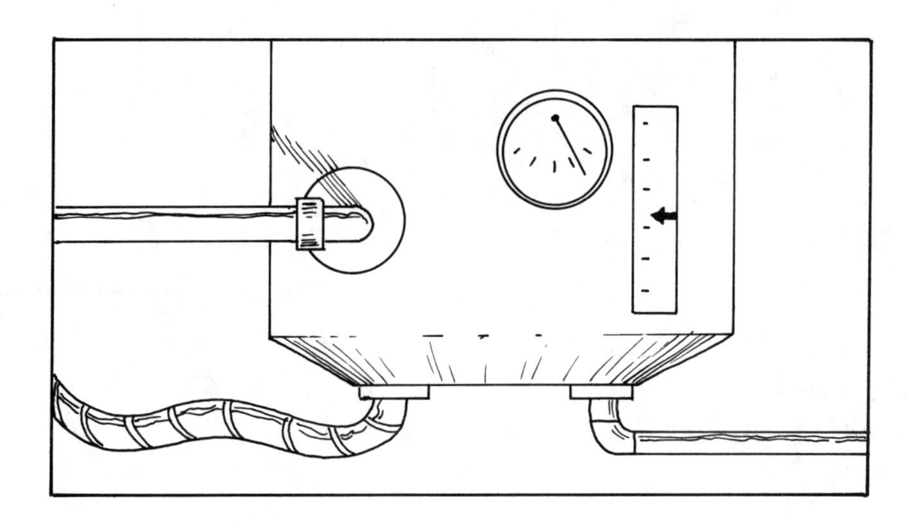

Fold

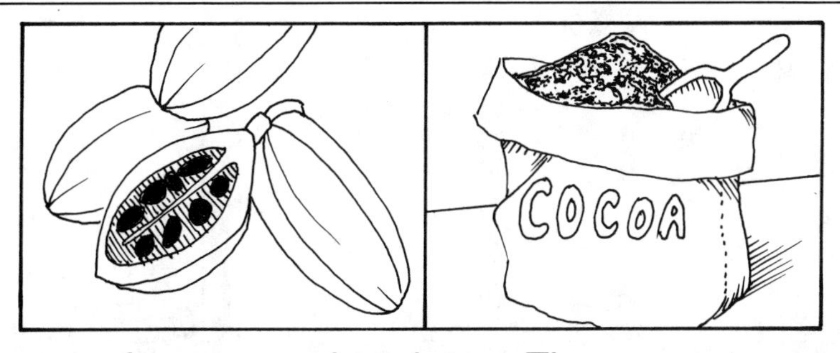

Chocolate comes from a seed pod, too. The cacao tree grows in tropical parts of the world. The ripe seed pods are split open and the beans are scooped out. The cacao beans are put into boxes to ferment for several days. Then the beans are dried and put in sacks to be sent to the chocolate factories. At the factory the cacao beans are cleaned and roasted. The roasted beans are cracked. Inside the beans is pure chocolate. The chocolate is crushed into a brown liquid. Part of the liquid is removed leaving the chocolate that is made into cocoa powder. This powdered cocoa is put into bags and sent to the ice cream factory.

Think About It:

How many different ways have you eaten chocolate? Was the chocolate you used a liquid or a powder?

Circle your favorite way to have ice cream.

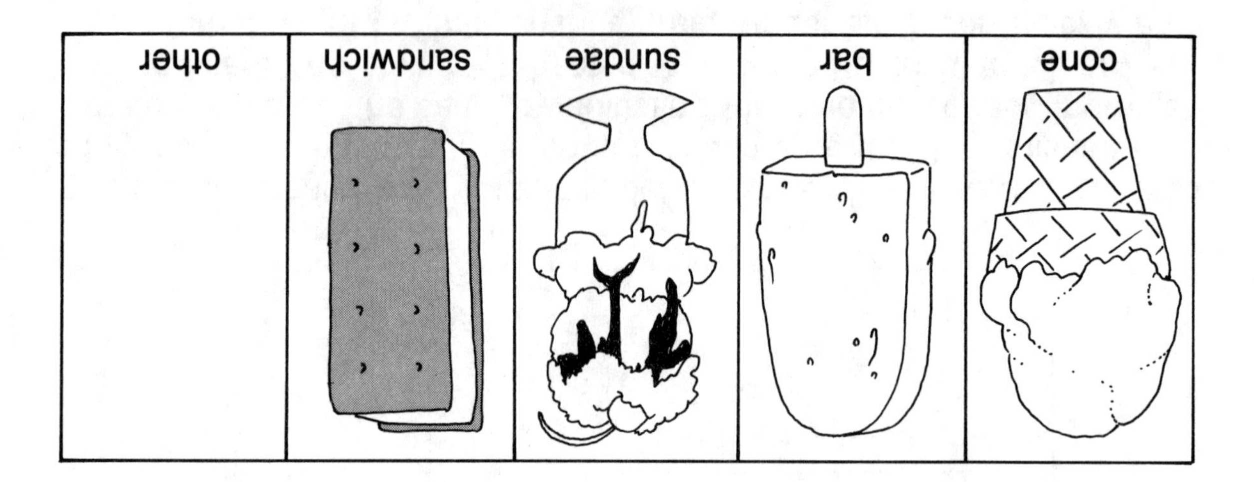

Ice cream is packaged in many different ways. This means you can buy ice cream by the gallon for a big family or for a party. You can also buy ice cream packaged in individual servings like ice cream bars or sandwiches.

Fold

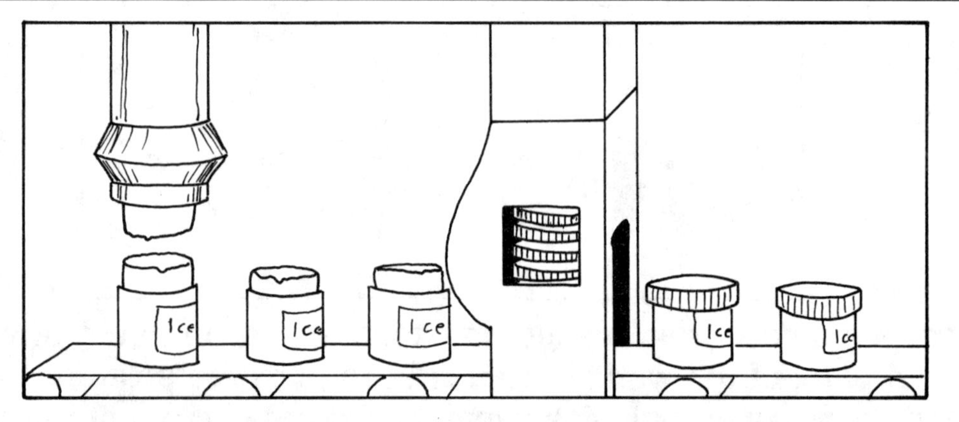

Flavorings, fruits and nuts are added to the thick ice cream paste. The ice cream is pressed through a round or square tube. As it comes out of the tube it is cut into pieces and put into containers. The ice cream is frozen solid and stored in a huge freezer at 20 degrees below zero. It is delivered to stores in special trucks that are like freezers on wheels so the ice cream is fresh and firm when you buy it.

Think About It:

How many different people earn money each time a container of ice cream is sold?

Note: These questions can be used in a whole class discussion or in cooperative-learning groups.

Think About Ice Cream

How many different flavors can you list?

Which of your senses do you use to determine ________________?

the flavor of the ice cream
if it is frozen solid or melted
if it is plain or has added ingredients
if it is smooth or grainy

List other frozen desserts.

Which is healthier for you...Why?

ice cream or ice milk
ice cream or frozen yoghurt
natural ingredients or man-made ingredients
ice cream or fruit

What other foods are made from milk or cream?

Label your container of ice cream.

A New Ice Cream Sensation

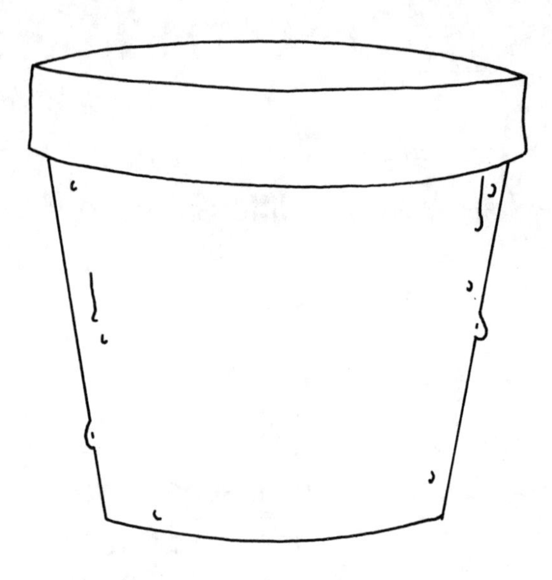

Invent a new and unusual ice cream flavor.

1. What is the name of your ice cream?

____

____

2. What special ingredients does it contain?

____

____

3. What does it look like?

____

____

4. What is the best way to serve your ice cream?

____

____

5. Where would you sell it?

____

____

6. Why should anyone buy this new ice-cream flavor?

____

____

Note: Use this activity to practice reading a chart for information needed to answer a question. Put the chart on the chalkboard or on a sheet of butcher paper. Questions may be answered in oral or written form.

BANANA SPLITS!

item	price
ice cream	60 cents per scoop
banana	20 cents per slice
chocolate sauce	15 cents
whipping cream	24 cents
walnuts	18 cents
cherry	10 cents each

What would these banana splits cost?

1 scoop ice cream
chocolate sauce
2 slices banana
1 cherry

____

1 scoop ice cream
1 slice banana
whipping cream
walnuts

____

2 scoops ice cream
2 banana slices
chocolate sauce
whipping cream

____

3 scoops ice cream
2 slices banana
whipping cream
3 cherries

____

2 scoops ice cream
2 slices banana
chocolate sauce
2 cherries

____

1 scoop ice cream
1 slice banana
chocolate sauce
walnuts
whipping cream
1 cherry

____

Create a banana split for yourself.

What will it contain? ________________________________

How much will it cost? ________________________________

Note: Brainstorm to create lists of good descriptive words before you turn your "poets" loose to write their cinquains. Do several examples with your students if this is a new experience for them. Have them write their finished poems on the ice-cream pattern on the following page.

Ice Cream Poets

Follow these steps to write a cinquain about your favorite flavor of ice cream.

1. Write one word naming the flavor. ______________________

2. Write two words describing the color.

   ______________________ ______________________

3. Write three words describing the texture.

   ______________ ______________ ______________

4. Write four words describing the taste.

   __________ __________ __________ __________

5. Write one word referring back to the flavor. ______________

Examples:

Vanilla
Snowy white
Smooth as silk
Sweet lumps of flavor
Creamy

Neapolitan
Many colors
Soft — no lumps
Yummy blend of tastes
Cold

Copy your finished poem on the ice-cream cone pattern. Cut a paper scoop of ice cream to paste to the top of the cone.

Note: Reproduce this pattern on light brown construction paper. It can be used with both the poetry and riddle writing activities in this unit.

Ice-Cream Cone Pattern

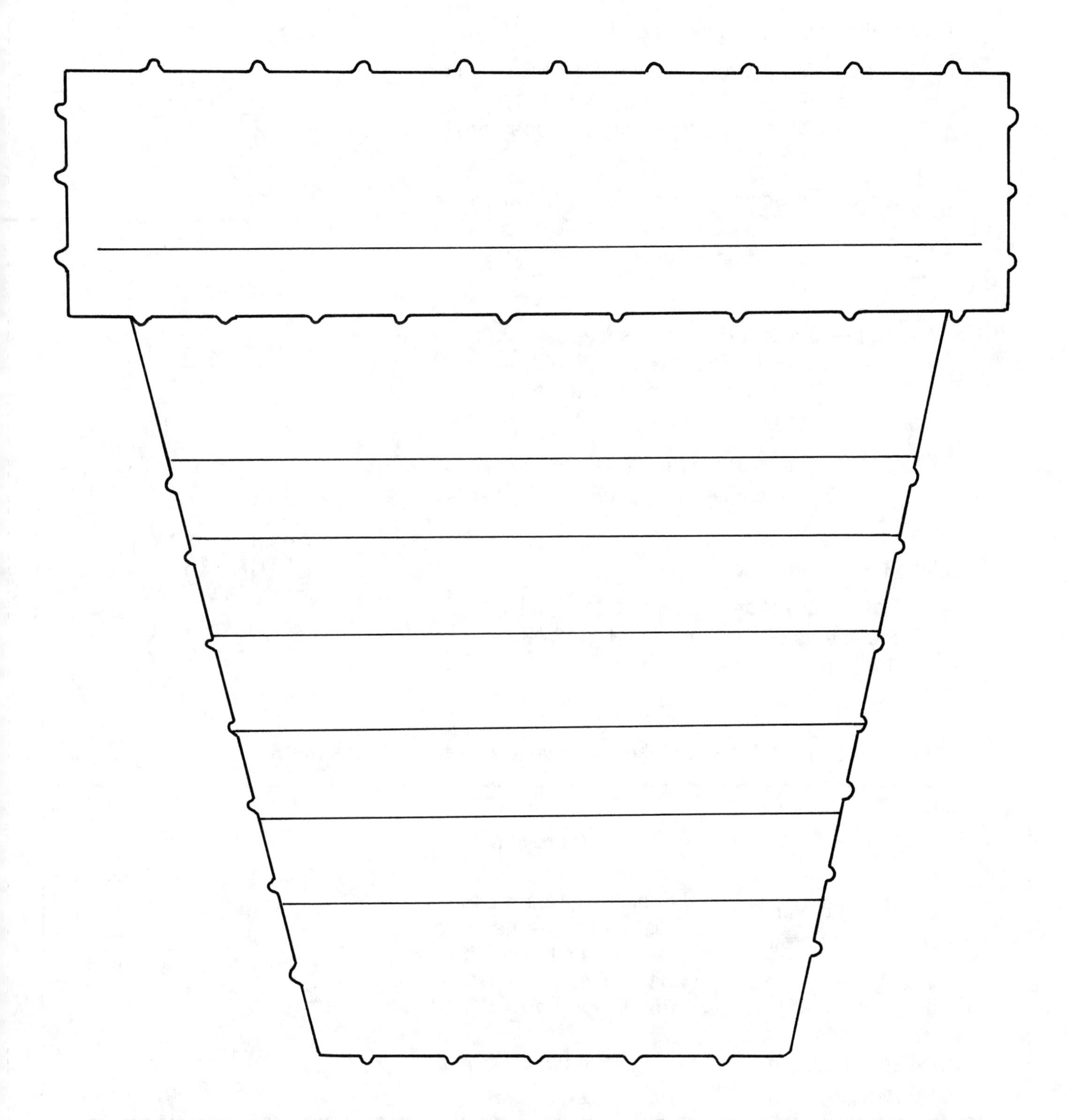

Note: Each child will need a copy of the ice cream-cone pattern on page 29, a 5" (13 cm) square of construction paper, and an 8½" X 11" (20.3 X 28 cm) sheet of tag.

An Accordion-Flap Book

Ice-Cream Riddles

Follow these directions to create a charming accordion book to display riddles written by your students.

Create the ice-cream cone riddle:

Cone — Have each child write a riddle about a flavor (real or imaginary) of ice cream on the ice-cream cone pattern.

Ice-Cream Scoop — Reproduce the pattern on the inside back cover. Fold on the dotted line. Paste the flap above the cone.

Tag — Children paste the cone to the tag. Make a fold in one side of the ice-cream scoop and paste it to the tag to create a flap.

Lift the flap and write the answer to the riddle underneath.

Make the accordion book:

Make a cover on the same size tag. Tape the sheets of tag together. (Tape both the front and back sides.)

Display the accordion book where everyone can try to guess the riddles.

Examples:

(simple)

I'm brown and creamy
and full of chocolate bits.

What am I?

Chocolate Chocolate Chip Ice Cream

(in rhyme)

I'm pink with hard bits,
From a cane Santa brings,
I'm striped all over with red and white rings.
What flavor am I?

Peppermint Candy Ice Cream

(strange)

I'm full of little fishes and red with tomato sauce.
You will taste yummy cheesy lumps as I melt in your mouth.

What am I?

Anchovy Pizza Ice Cream

Note: Use this board to display original writings about ice cream by your students.

A Bulletin Board

1. Cover the chalkboard with blue butcher paper.

2. Cut large cones from brown construction paper. Add lines with brown or black marking pen.

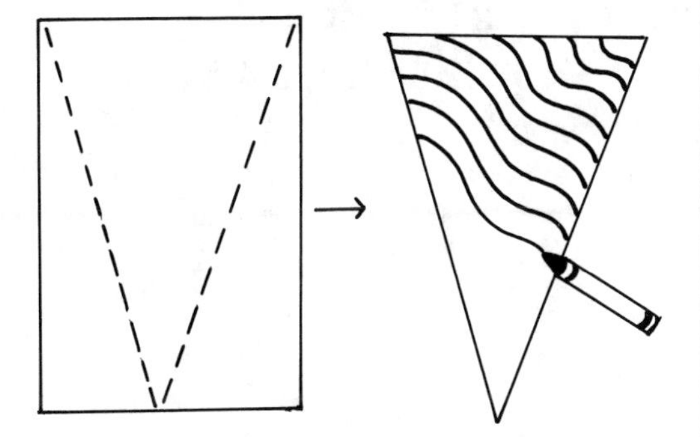

3. Cut large circles in various colors to be the scoops of ice cream. Pin a different child's paper to each cone.

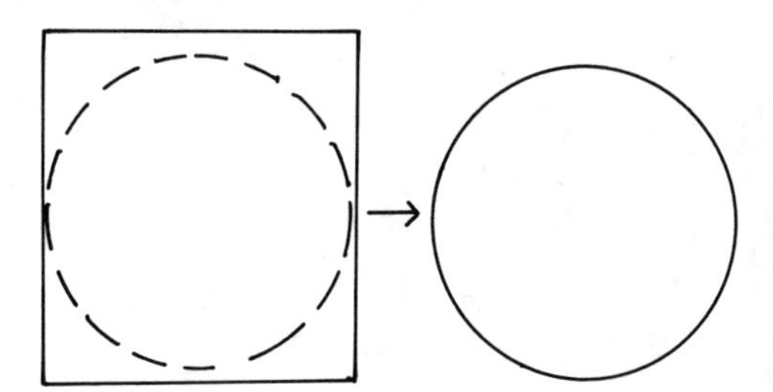

Display:

Original poems about ice cream.
Facts about ice cream and its ingredients.
Steps in making ice cream.
The new flavors "invented" by your students.

Let's Eat

Follow these steps to create three-dimensional ice-cream "eaters."

Materials:
drawing paper 9" X 12" (22.8 X 30.5 cm)
brown construction paper —
2½" X 3½" (6.35 X 8.9 cm)
white construction paper —
1" X 6" (2.54 X 15.24 cm)
cotton balls
crayons
scissors
paste or glue

Follow these steps:

Cone

1. Cut brown construction paper as shown. Draw lines with a brown or black crayon.
2. Form cone and paste together. Add cotton ball to the top.

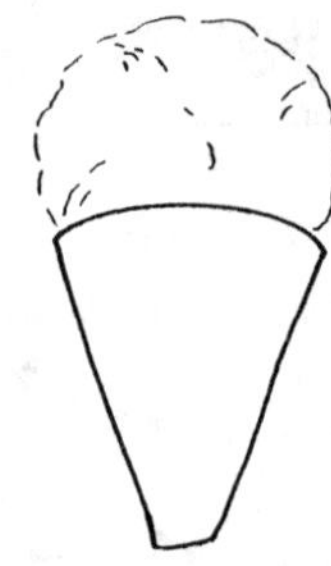

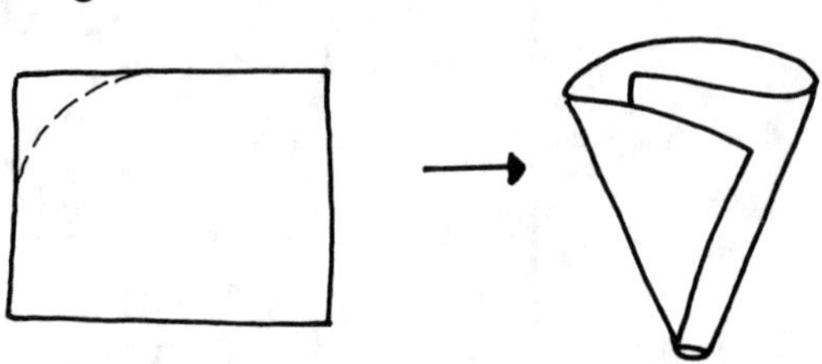

Drawing

1. Paste the cone in the center of the drawing paper.
2. Draw yourself eating the cone.

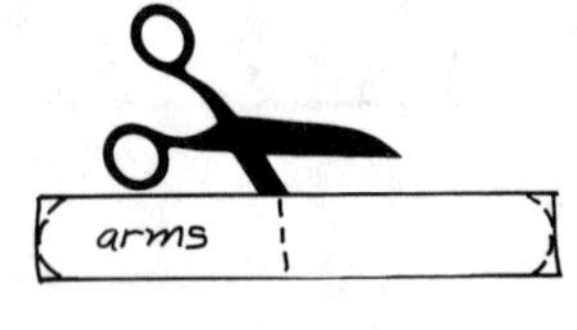

3. Cut hands from a scrap of white paper. Color the hands and paste to the picture as if they are holding the cone.

"21 Flavor" Ice Cream Search

P	E	P	P	E	R	M	I	N	T	S	T	I	C	K	I
I	X	B	U	T	T	E	R	P	E	C	A	N	O	B	C
S	V	A	N	I	L	L	A	X	I	C	E	B	O	L	E
T	S	Q	P	I	N	E	A	P	P	L	E	U	K	A	P
A	T	B	C	H	O	C	O	L	A	T	E	B	I	C	E
C	R	A	R	A	S	P	B	E	R	R	Y	B	E	K	A
H	A	N	E	A	P	O	L	I	T	A	N	L	S	W	N
I	W	A	C	R	E	A	M	I	C	E	Y	E	A	A	U
O	B	N	R	X	A	I	C	E	H	X	U	G	N	L	T
C	E	A	E	I	C	E	C	R	E	A	M	U	D	N	B
O	R	N	A	X	H	I	C	E	R	X	Q	M	C	U	U
F	R	U	M	R	O	C	K	Y	R	O	A	D	R	T	T
F	Y	T	I	N	R	O	O	F	Y	I	C	E	E	T	T
E	F	R	E	N	C	H	V	A	N	I	L	L	A	C	E
E	L	E	M	O	N	C	U	S	T	A	R	D	M	E	R

BANANA NUT
BLACK WALNUT
BUBBLE GUM
BUTTER PECAN
CHERRY
CHOCOLATE
COFFEE
COOKIES AND CREAM
FRENCH VANILLA
LEMON CUSTARD
NEAPOLITAN
PEACH
PEANUT BUTTER
PEPPERMINT STICK
PINEAPPLE
PISTACHIO
RASPBERRY
STRAWBERRY
ROCKY ROAD
TIN ROOF
VANILLA

Try to find ICE CREAM in this word search. Circle it in red.

Other Ice-Cream Activities

"School-made" Ice Cream — Making ice cream in class is educational as well as fun. Children have an opportunity to practice reading and following directions as well as practicing measurement skills.

Home-made vs Store-bought — Compare one or more brands of purchased ice cream with homemade.

a. Check the temperature of store-bought and homemade.

b. Time the melting rate.

c. Have a taste test.

What does the label say? — Read and compare the labels of various brands.

a. contents

b. calorie count

How many scoops? — Measure the number of standard scoops in containers of various sizes.

a. pint

b. quart

c. half gallon

Ice-Cream Graphs — Make bar or line graphs comparing such topics as...

a. favorite flavors

b. homemade vs store-bought

c. favorite brands

d. favorite form (cone, bar, sandwich, dish)

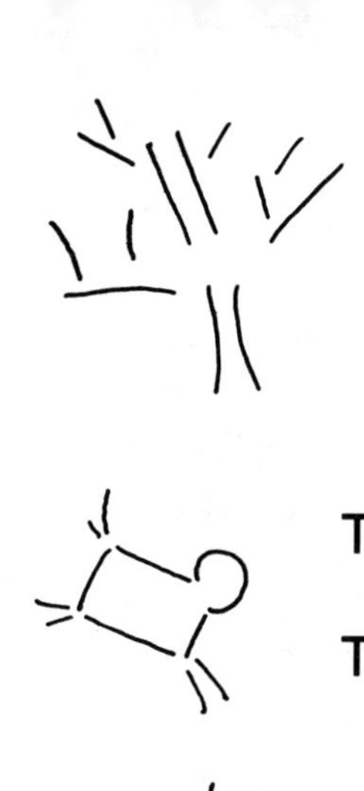

Pencils

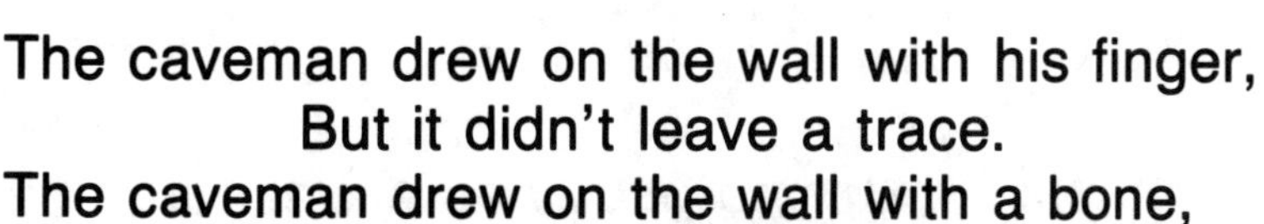

The caveman drew on the wall with his finger,
But it didn't leave a trace.
The caveman drew on the wall with a bone,
But it remained an empty space.

The caveman drew on the wall with a stick,
One that he plucked from the fire.
He filled up that space with trace upon trace
Until he couldn't reach higher.

I use my pencil to write stories
About fierce dragons and brave knights.
I then use my pencil to draw pictures,
Showing castles and medieval sights.

Some kids chew on their pencils.
Others use them to scratch their heads.
Some make drum beats with their erasers.
I like to write with my pencil instead.

by Leslie Tryon

Books to Read About Pencils:

From Graphite to Pencil by Ali Mitgutsch; Carolrhoda Books, Inc., 1982

"Pencils" World Book Encyclopedia, Volume 15, 1987 Edition, pages 209-210

"Pencils" Childcraft, The How and Why Library Volume 7, pages 122-123

Note: Reproduce pages 36, 37, and 38. Fold the pages in half and staple them together on the left side to make a book to take home.

The materials for a pencil come from many different places.

The wood for the outside of the pencil is made from the trunks of cedar trees. The trees are cut down by loggers. The branches are cut off and the trunks are taken to a mill where they are cut into narrow pieces which are sent to the pencil factory.

We call the part of the pencil that writes 'lead', but it is not made of lead. It is a soft form of carbon called graphite and clay. The graphite is taken from a mine and sent to the pencil factory.

The pencil factory will also need such materials as paint, metal, and rubber to finish the pencil. Once the factory has all of the materials, it can begin to make the parts and put them together.

Think About It:

Look at this pencil. Name what each part is made of.

2

Fold

How a Pencil is Made

Do you ever think about your pencil? You probably use it every day and only think about it when the lead breaks or you cannot find it. A pencil only has a few parts, but each part is important. Let's look at how a pencil is put together.

Think About It:

Look at your own pencil. Name the different parts.

lead	eraser
wood casing	ferrule

1

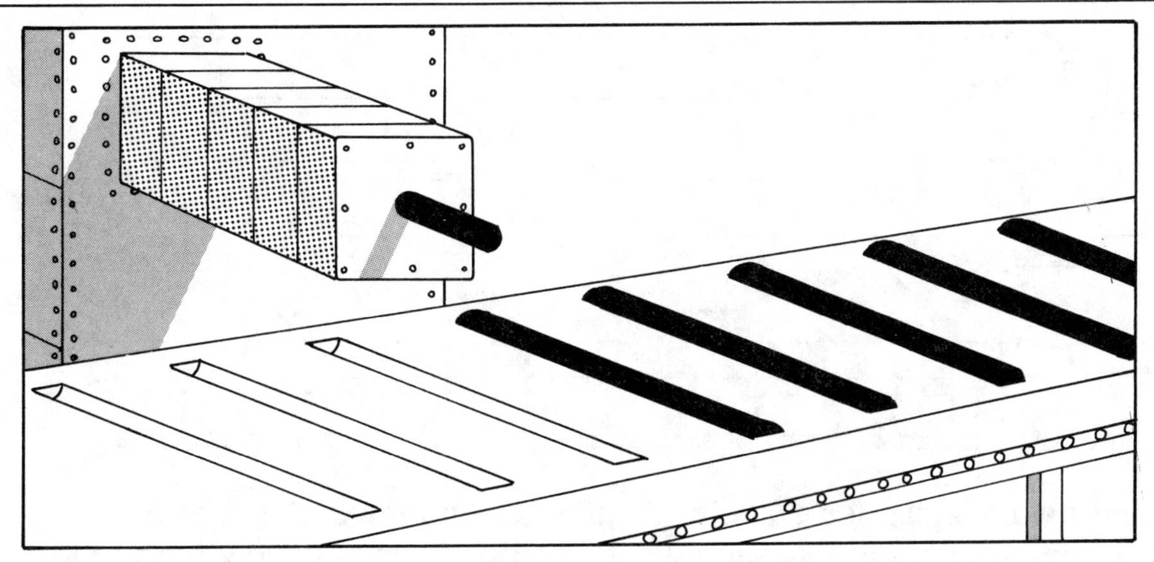

Special machines mix the graphite with water and clay. The mixture is packed in a cylinder and pushed through a small hole to make a long strand of pencil lead. The long piece is cut to make pieces the right size for a pencil. The lead is baked in a hot oven to make it hard. It is dipped in wax to make it smooth.

Think About It:

Measure a brand new pencil. How long is the lead piece?

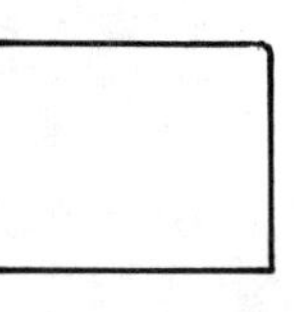

3

Fold

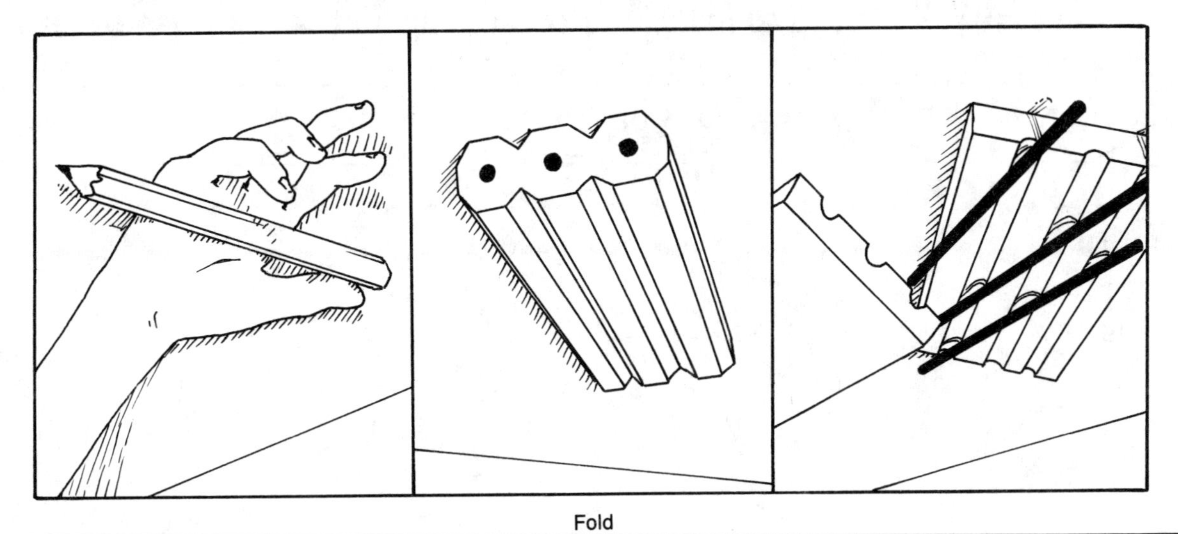

Other machines are used to make the outside of the pencil. The cedar slats are cut the same length as a pencil. Round groves are cut into the slats. The lead pieces are put into the groves. Two slats are glued together. Clamps hold the slats until they dry. Then slats are cut into the shape of a pencil.

Think About It:

Why do you think most pencils are cut so they have six sides?

4

The eraser on your pencil is made of a mixture of things. There is rubber of course, but it also contains vegetable oil, fine pumice, and sulfur. The eraser is attached to the pencil with a piece of metal called a ferrule. This holds the eraser to the pencil. The finished pencils are put into boxes and sent in delivery trucks to many different types of stores. The next time you buy a pencil, think about all of the people, all of the machinery, and all of the work it took to make it.

Think About It:

Name all of the places pencils are sold.

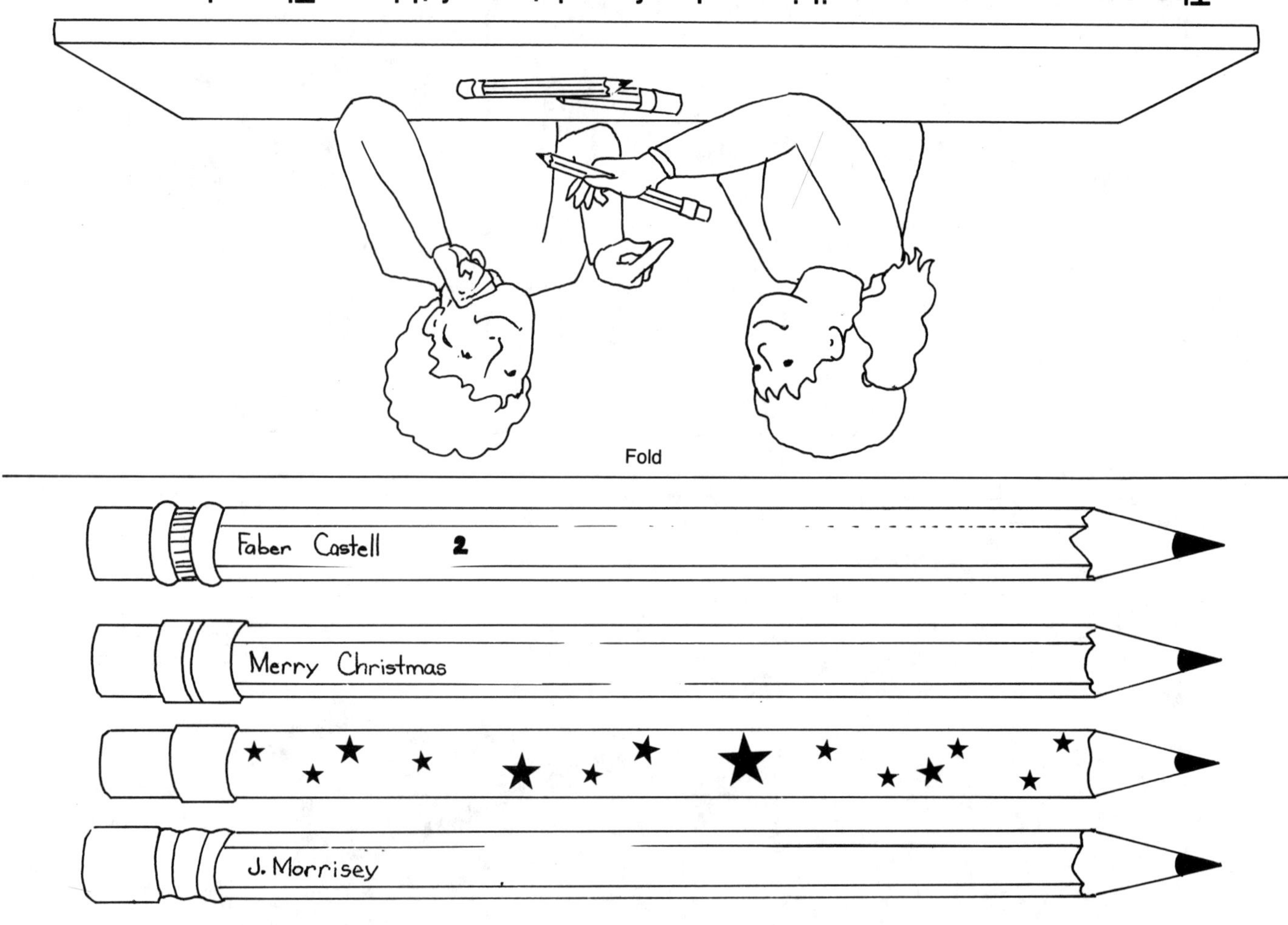

The pencil is sanded smooth, painted and trimmed. A name and number are pressed into the wood. The number tells how hard the lead is. You are probably using a number 2 pencil. The name on the pencil is usually the name of the pencil manufacturer. Sometimes you will get a pencil with a special message written on it.

Think About It:

Look at your pencil to find the name of the manufacturer.

Get several pencils with different numbers. See if you can tell which are soft and which are hard.

Note: These questions can be used in a whole class discussion or in cooperative-learning groups.

Think About Pencils

How many uses can you think of for a pencil?

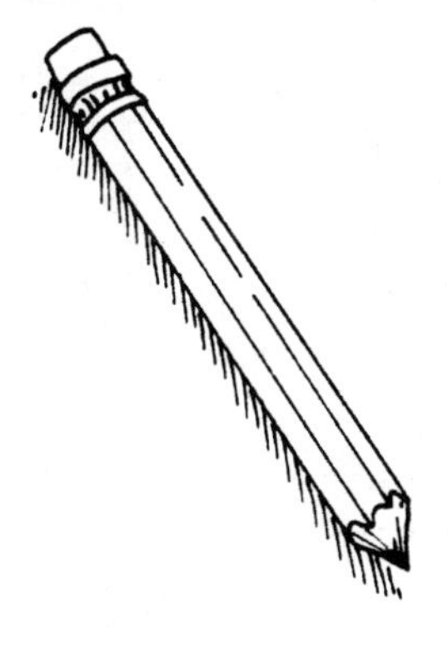

Why are pencils so widely used?

Name other items you can write with.
Why is writing important?

Pretend that all pencils, pens, typewriters, etc. have disappeared.
Invent a new writing instrument.

Whose Pencil is This?

Use this code key to help you discover whose pencils these are.

a-z	e-v	i-r	m-n	q-j	u-f	y-b
b-y	f-u	j-q	n-m	r-i	v-e	z-a
c-x	g-t	k-p	o-l	s-h	w-d	
d-w	h-s	l-o	p-k	t-g	x-c	

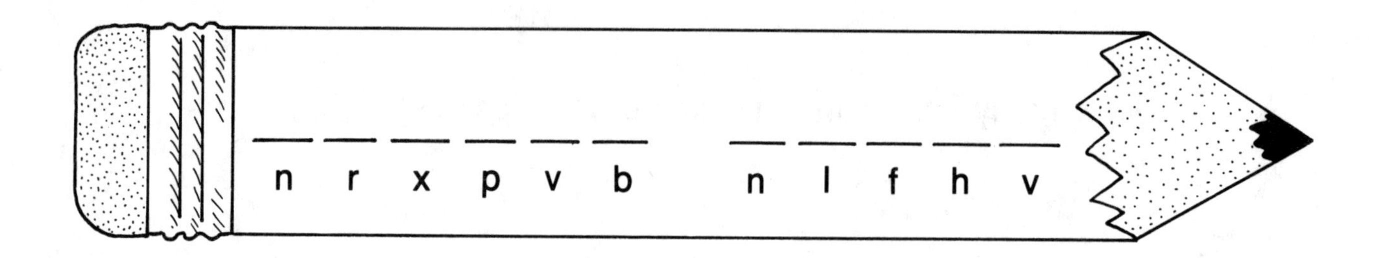

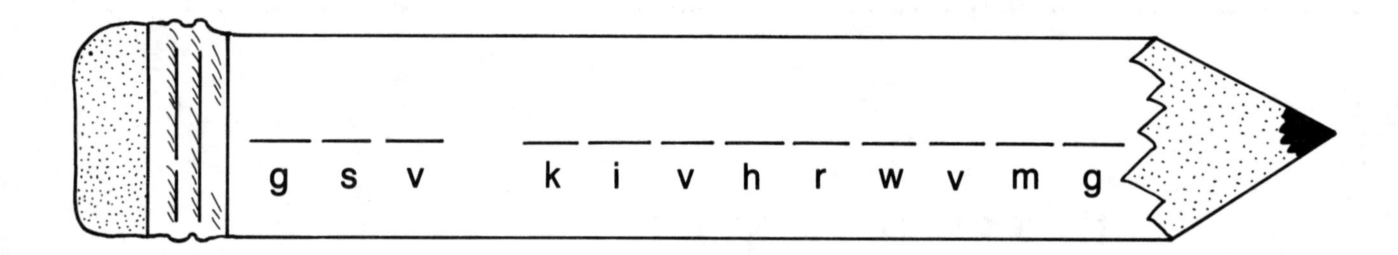

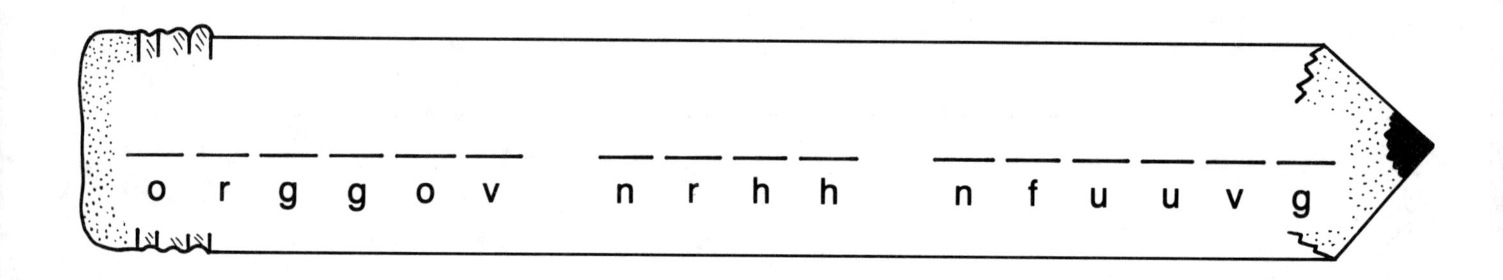

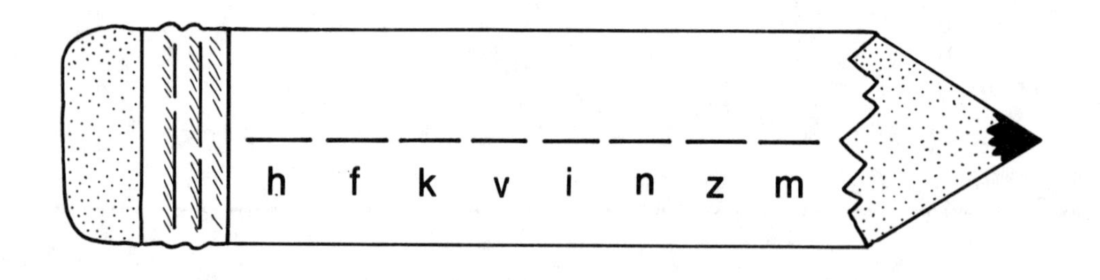

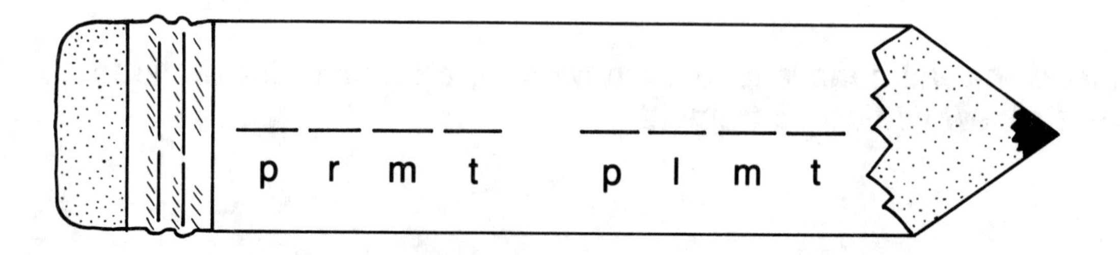

Find the Missing Pencils

Harry was running down the path when he tripped over a rock. The box of pencils he was carrying spilled and pencils flew all over the place. See if you can find all of Harry's lost pencils.

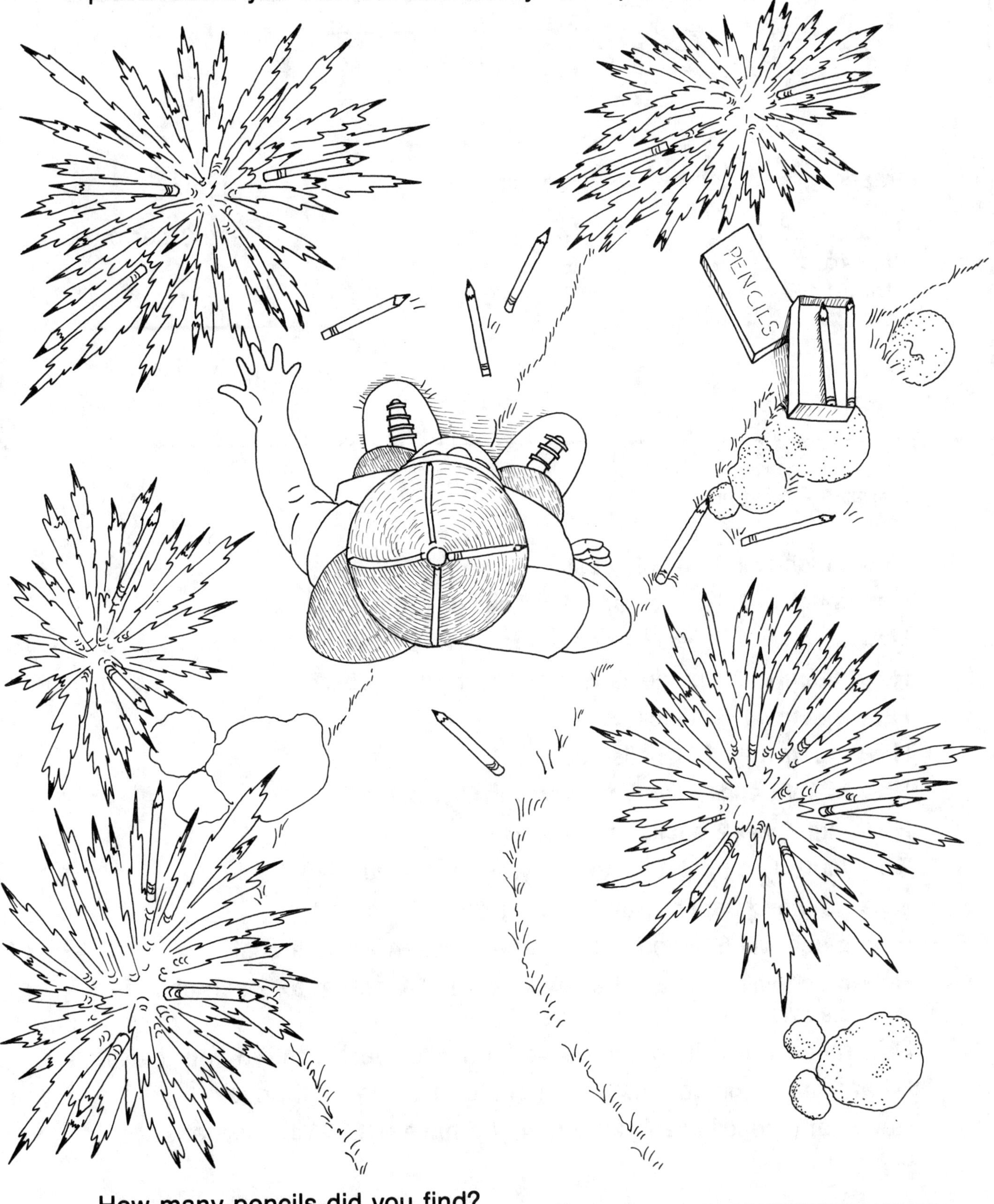

How many pencils did you find? ______________________

Note: Make this chart on the chalkboard or chart paper. Select questions appropriate for the grade level and ability of your students. Encourage them to make up questions to ask their classmates. These questions may be answered orally or in writing.

Hurry to the Great Pencil Sale!

eraser

25 cents each
2 for 45 cents
6 for $1.25

pencil

14 cents each
2 for 25 cents
6 for 70 cents

paper

$1.50 per pack

binder

$3.25 each

Sample Questions:

How much will 3 pencils cost?
How much will 1 pencil and 1 eraser cost?
How much will 2 pencils and 1 eraser cost?
How much will 1 binder and 1 pack of paper cost?
How much will 5 pencils cost?
How much will 12 pencils cost?
How many pencils can you buy with fifty cents?
How much will 2 packs of paper cost?
Can you buy a pack of paper if you have 4 quarters?
What is the cheapest way to buy a dozen pencils?
How much will 5 dozen pencils and 3 dozen erasers cost?
How much will it cost to buy 1 pencil and 1 eraser for each child in your class?
How much would it cost to buy a binder for each child in your class?
What would a binder cost if you could buy it for half price?
How many pencils can you buy if you have $10? Will you get any change back?

Note: Each child will need a copy of the form on the following page and a 3" X 7½" (7.6 X 19 cm) strip of construction paper.The form can also be used for couplets about pencils or for pencil riddles.

What Does My Pencil Look Like?
A Pop-Up Pencil Activity

Guide your students through these steps to create their pencil drawing, descriptive paragraph, and pop-up form.

Pencil — Have children think about the type of pencil they are going to make. Is it new or old? Does it have a special pattern or message printed on it? Does its condition tell a story?

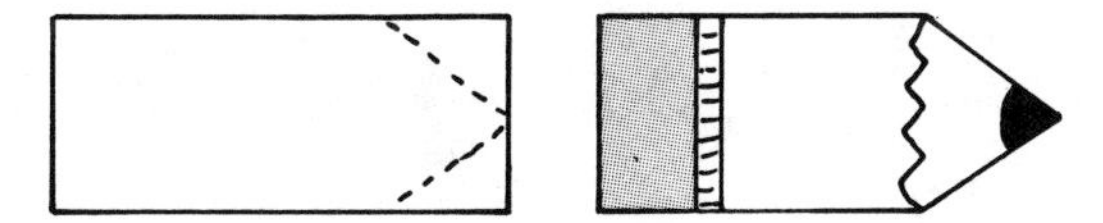

Draw pencil details on the strip of construction paper. Cut off the tip to form the pencil point.

Descriptive Paragraph — Have each child write a paragraph describing the pencil he/she has created. The finished paragraph should be copied on the lines at the bottom of the pop-up form.

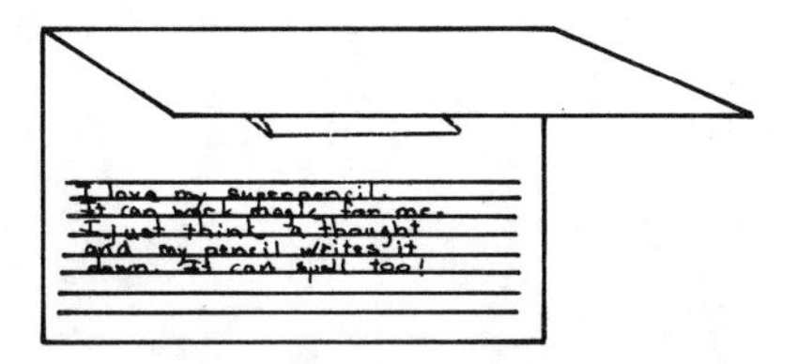

Pop-Up Form — Guide students through these steps to put the pop-up form together.

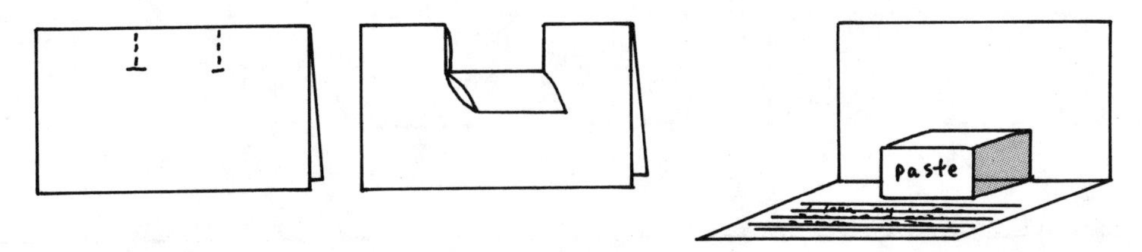

These pop-up pencils can be put into a construction paper folder. Fold a sheet of construction paper (slightly larger than the pop-up form) in half to make a cover, then paste as shown below.

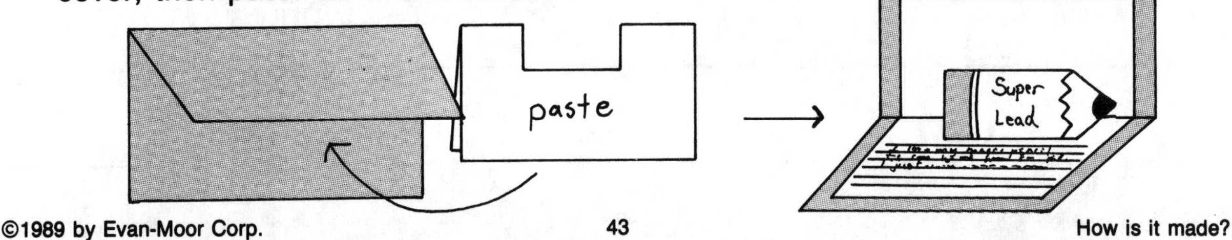

cut

cut

fold

Note: Run the patterns below on construction paper or tag. Cut the slits with an Exacto knife. Provide scraps of paper and crayons or marking pens for children to use in creating original pencil pals. Slip the completed "pal" on the end of a pencil.

Pencil Pals

Color, cut out, and slip on your pencil.

Use these forms to make your own "pencil pals."

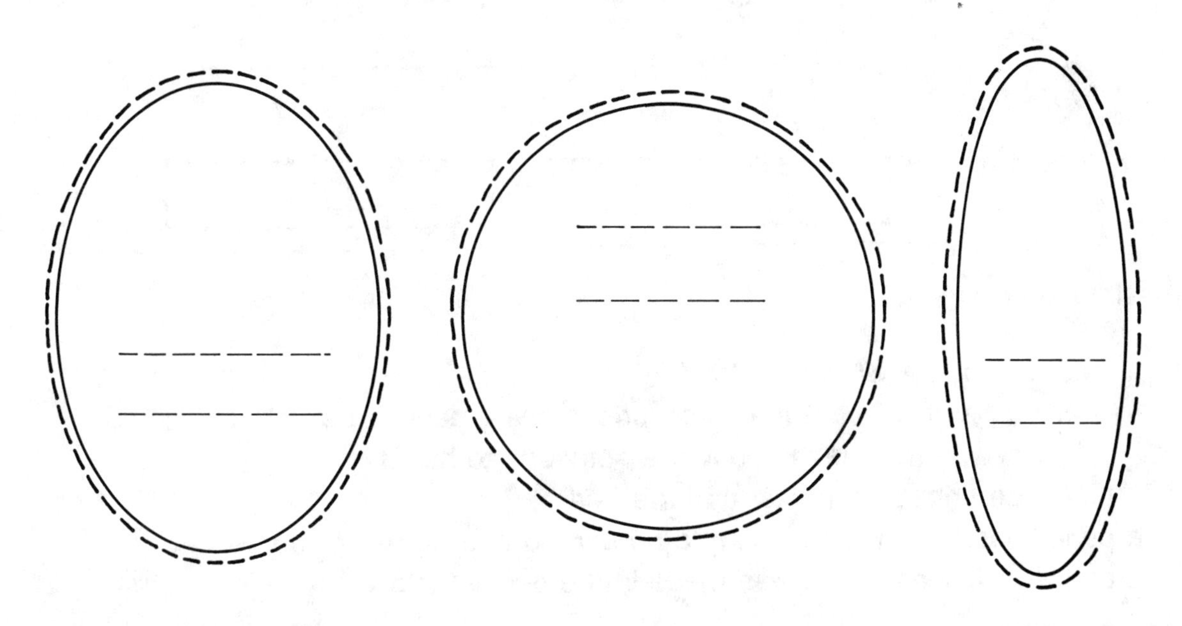

Note: Reproduce these pencils. Have your students select one of the pencils to write about. Attach the pencil to a sheet of writing paper. Have the story told from the pencil's point of view.

Conversation With a Pencil

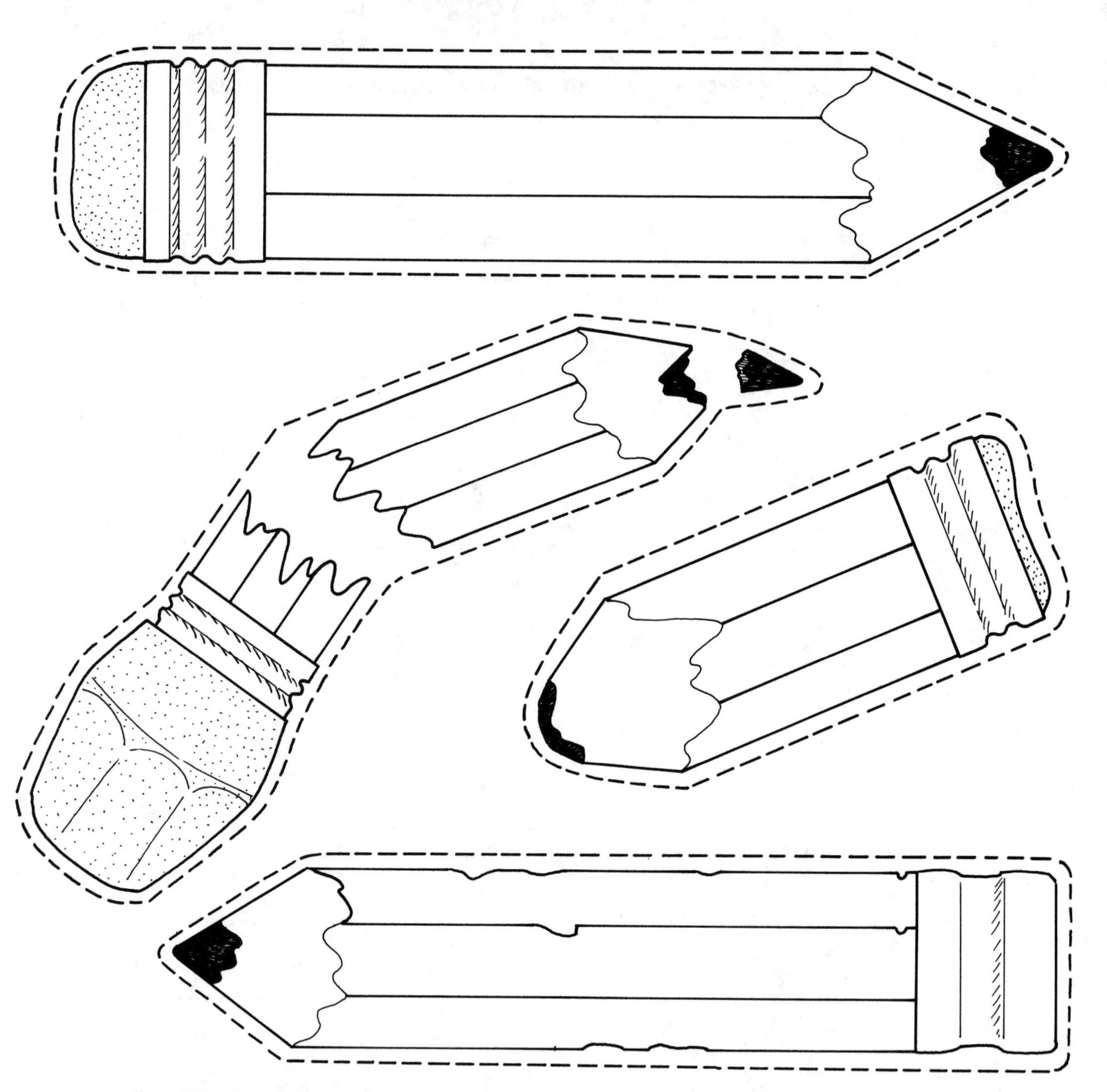

Possible Topics to Use:

an old, stubby pencil recalls its life
a pencil covered in tooth marks describes being pencil to a "pencil biter"
a shiny, new pencil tries to convince someone to buy it
a pencil describes being lost under a bed
a pencil describes what it's like to be carried in a pocket or purse
a pencil describes how it was made in the pencil factory

A Bulletin Board

1. Cover the bulletin board with light blue butcher paper.

2. Make a giant pencil from butcher or construction paper. Add details with black marking pen. Add "pencil marks" with black roving.

3. Pin up pictures your students have made using only pencils as drawing tools.

Variations:

Write On... — Display original stories or samples of good handwriting.

How a Pencil is Made — Sequence the steps in a the construction of a pencil.

Pencil Poems — Display original poems about pencils.

History of the Pencil — Display sentence strips containing facts about the history of the pencil.

Other Pencil Activities

Pencil Pictures — Provide a supply of pencils of various hardness for children to use in creating pictures.

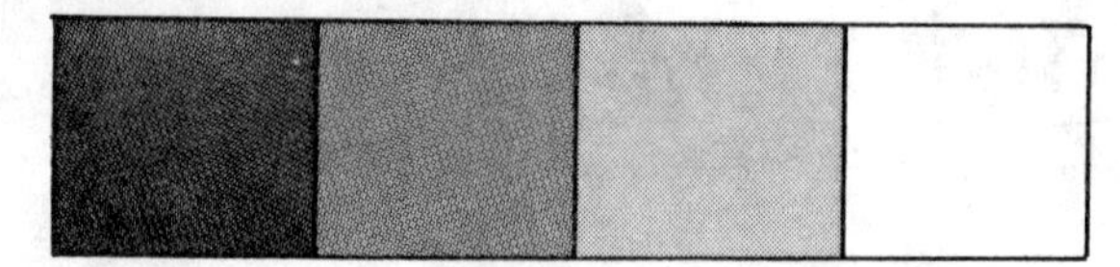

1. Look at books containing pencil illustrations (books by Chris Van Allsburg for example).
2. Practice shading from light to dark using only pencils.
3. Draw original pictures using one or more pencils.
4. Display the pictures on a bulletin board.

Pencil Poems — Have your students write couplets about pencils, erasers, writing, etc. Remind them that a couplet is two lines that rhyme, with the same number of syllables in each line. (Don't worry about exact syllable count with second graders.)

My pencil is brand new.
Its color is bright blue.

I was taking a test when my pencil broke.
Now I'm in big trouble and that is no joke.

Why does a pencil eraser always wear out
While the lead is still long as an elephant's snout?

Pencil Search — Practice locating information in the dictionary, encyclopedia, or in library books. Children may work independently or in cooperative-learning groups.

1. What does it mean? — Find these words in the dictionary.

    a. lead
    b. graphite
    c. pencil
    d. ferrule

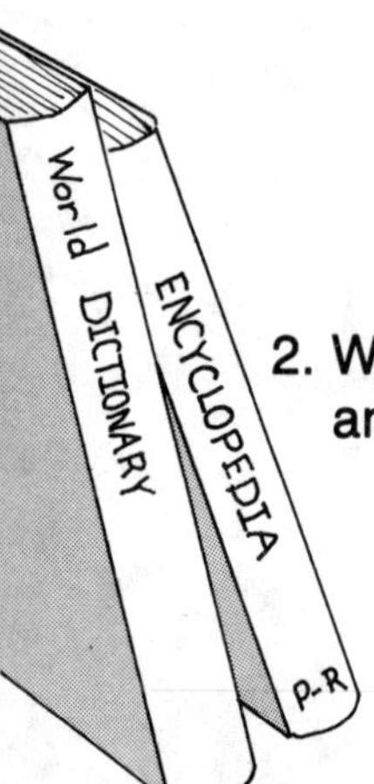

2. When was it invented? — Use the encyclopedia or library books to locate the answers to questions about the history of the pencil. For example:

    a. When were the first pencils used?
    b. Who made the first pencils in the U.S.A.? Where? When?
    c. When did writing become necessary?